Blockchain: Applications, Challenges, and Solutions

Blockchain: Applications, Challenges, and Solutions

Editors

Ahad ZareRavasan
Taha Mansouri
Michal Krčál
Saeed Rouhani

MDPI • Basel • Beijing • Wuhan • Barcelona • Belgrade • Manchester • Tokyo • Cluj • Tianjin

Editors
Ahad ZareRavasan
Department of Corporate
Economy, Faculty of
Economics and
Administration
Masaryk University,
Brno, Czech Republic

Taha Mansouri
School of Science,
Engineering & Environment,
the University of Salford,
Salford, Greater Manchester, UK

Michal Krčál
Department of Corporate
Economy, Faculty of
Economics and
Administration Masaryk University,
Brno, Czech Republic

Saeed Rouhani
Faculty of Management,
University of Tehran,
Tehran, Iran

Editorial Office
MDPI
St. Alban-Anlage 66
4052 Basel, Switzerland

This is a reprint of articles from the Special Issue published online in the open access journal *Future Internet* (ISSN 1999-5903) (available at: https://www.mdpi.com/journal/futureinternet/special_issues/B_ACS).

For citation purposes, cite each article independently as indicated on the article page online and as indicated below:

LastName, A.A.; LastName, B.B.; LastName, C.C. Article Title. *Journal Name* **Year**, *Volume Number*, Page Range.

ISBN 978-3-0365-6473-9 (Hbk)
ISBN 978-3-0365-6474-6 (PDF)

© 2023 by the authors. Articles in this book are Open Access and distributed under the Creative Commons Attribution (CC BY) license, which allows users to download, copy and build upon published articles, as long as the author and publisher are properly credited, which ensures maximum dissemination and a wider impact of our publications.
The book as a whole is distributed by MDPI under the terms and conditions of the Creative Commons license CC BY-NC-ND.

Contents

Ahad ZareRavasan, Taha Mansouri, Michal Krčál and Saeed Rouhani
Editorial for the Special Issue on Blockchain: Applications, Challenges, and Solutions
Reprinted from: *Future Internet* **2022**, *14*, 155, doi:10.3390/fi14050155 . . . . **1**

Geneci da Silva Ribeiro Rocha, Letícia de Oliveira and Edson Talamini
Blockchain Applications in Agribusiness: A Systematic Review
Reprinted from: *Future Internet* **2021**, *13*, 95, doi:10.3390/fi13040095 . . . . **3**

Claudia Antal, Tudor Cioara, Ionut Anghel, Marcel Antal and Ioan Salomie
Distributed Ledger Technology Review and Decentralized Applications Development Guidelines
Reprinted from: *Future Internet* **2021**, *13*, 62, doi:10.3390/fi13030062 . . . . **19**

Benjamin Leiding, Priyanka Sharma and Alexander Norta
The Machine-to-Everything (M2X) Economy: Business Enactments, Collaborations, and e-Governance
Reprinted from: *Future Internet* **2021**, *13*, 319, doi:10.3390/fi13120319 . . . . **51**

Evgenia Kapassa and Marinos Themistocleous
Blockchain Technology Applied in IoV Demand Response Management: A Systematic Literature Review
Reprinted from: *Future Internet* **2022**, *14*, 136, doi:10.3390/fi14050136 . . . . **67**

Haoli Sun, Bingfeng Pi, Jun Sun, Takeshi Miyamae and Masanobu Morinaga
SASLedger: A Secured, Accelerated Scalable Storage Solution for Distributed Ledger Systems
Reprinted from: *Future Internet* **2021**, *13*, 310, doi:10.3390/fi13120310 . . . . **87**

Nur Arifin Akbar, Amgad Muneer, Narmine ElHakim and Suliman Mohamed Fati
Distributed Hybrid Double-Spending Attack Prevention Mechanism for Proof-of-Work and Proof-of-Stake Blockchain Consensuses
Reprinted from: *Future Internet* **2021**, *13*, 285, doi:10.3390/fi13110285 . . . . **107**

Ronghua Xu, Deeraj Nagothu, Yu Chen
EconLedger: A Proof-of-ENF Consensus Based Lightweight Distributed Ledger for IoVT Networks
Reprinted from: *Future Internet* **2021**, *13*, 248, doi:10.3390/fi13100248 . . . . **127**

Neo C. K. Yiu
Toward Blockchain-Enabled Supply Chain Anti-Counterfeiting and Traceability
Reprinted from: *Future Internet* **2021**, *13*, 86, doi:10.3390/fi13040086 . . . . **151**

Editorial

Editorial for the Special Issue on Blockchain: Applications, Challenges, and Solutions

Ahad ZareRavasan [1,*], Taha Mansouri [2], Michal Krčál [1] and Saeed Rouhani [3]

1. Department of Corporate Economy, Faculty of Economics and Administration, Masaryk University, 61137 Brno, Czech Republic; michal.krcal@mail.muni.cz
2. School of Science, Engineering & Environment, the University of Salford, Salford M5 4WT, UK; t.mansouri@salford.ac.uk
3. Faculty of Management, University of Tehran, Tehran 14155-6311, Iran; srouhani@ut.ac.ir
* Correspondence: zare.ahad@mail.muni.cz

Blockchain is believed to have the potential to digitally transform and disrupt industry sectors such as finance, supply chain, healthcare, marketing, and entertainment. However, obstacles and challenges can be observed with its widespread applications. The Special Issue, "Blockchain: Applications, Challenges, and Solutions", in the Journal of Future Internet, covers the trending research topic of Blockchain applications, the challenges it faces, and the value it brings to different industry sectors. We received 15 submissions; nevertheless, after the initial screening and the peer review process, only eight papers have been finally accepted for publication. Accepted articles can be divided into two sets: (1) the review of applications and (2) technical solutions addressing the challenges of the technology.

The first set presents reviews of Blockchain applications in different domains. Rocha et al. [1] review blockchain applications in the agribusiness sector using a PRISMA-based systematic review. In 71 articles, they identified Blockchain applications for finance, energy, logistics, environmental, agricultural, livestock, and industrial support. They conclude that the research into blockchain applications in agribusiness is at an early stage, as most of the prototypes are in the developing or laboratory phase. Nevertheless, the applications could mature and promote greater reliability and agility in information with a reduced cost in the future. A comprehensive overview of Blockchain applications, challenges, solutions, alternatives, and usage for developing decentralized applications is presented in Antal et al. [2]. They employed a three-tier architecture for Blockchain applications to systematically classify the technology solutions. The paper presents a multi-step guideline for decentralizing the design and implementation of traditional systems. Leiding et al. [3] present the Machine-to-Everything (M2X) Economy concept, which follows an open, decentralized, and distributed smart-contract-based approach. M2X supports the corresponding multi-stakeholder ecosystem and facilitates M2X value exchange, collaborations, and business enactments. Kapassa and Themistocleous [4] use a systematic literature review (SLR) approach to analyze Blockchain applications in the area of Demand-Response Management (DRM) in the Internet of Vehicles (IoV). They end up with research challenges on blockchain-based DRM in IoV.

The second set of articles addresses technical solutions to the current challenges of Blockchain technology. Sun et al. [5] present an off-chain solution to relieve the storage burden of blockchain nodes while ensuring the integrity of the off-chain data. The solution is implemented based on Hyperledger Fabric (HLF). The authors' experimental results show that their solution significantly outperforms the original HLF. Akbar et al. [6] propose a hybrid algorithm that combines Proof-of-Stake (PoS) and Proof-of-Work (PoW) mechanisms to provide a fair mining reward to the miner/validator. The proposed algorithm can reduce the possibility of intruders performing double mining based on the

experimental results. Xu et al. [7] propose EconLedger, an Electrical Network Frequency (ENF)-based consensus mechanism that enables secure and lightweight distributed ledgers for small-scale Internet of Video Things (IoVT) edge networks. The proposed consensus mechanism relies on a novel Proof-of-ENF (PoENF) algorithm. A proof-of-concept prototype is developed and tested in a physical IoVT network environment. The experimental results on the designed prototype validate the feasibility of the proposed EconLedger to provide a trust-free and partially decentralized security infrastructure for IoVT edge networks. Finally, the key areas of decentralization, fundamental system requirements, and feasible mechanisms for developing decentralized product anti-counterfeiting and traceability ecosystems utilizing blockchain technology are identified in Yiu [8] via a series of security analyses compared with solutions currently implemented in the supply chain industry with centralized architecture.

We would like to thank all the authors for their papers submitted to this Special Issue. We would also like to acknowledge all the reviewers for their careful and timely reviews to help improve the quality of this Special Issue.

Funding: This research received no external funding.

Conflicts of Interest: The authors declare no conflict of interest.

References

1. Da Silva, R.R.G.; de Oliveira, L.; Talamini, E. Blockchain Applications in Agribusiness: A Systematic Review. *Future Internet* **2021**, *13*, 95. [CrossRef]
2. Antal, C.; Cioara, T.; Anghel, I.; Antal, M.; Salomie, I. Distributed Ledger Technology Review and Decentralized Applications Development Guidelines. *Future Internet* **2021**, *13*, 62. [CrossRef]
3. Leiding, B.; Sharma, P.; Norta, A. The Machine-to-Everything (M2X) Economy: Business Enactments, Collaborations, and e-Governance. *Future Internet* **2021**, *13*, 319. [CrossRef]
4. Kapassa, E.; Themistocleous, M. Blockchain Technology Applied in IoV Demand Response Management: A Systematic Literature Review. *Future Internet* **2022**, *14*, 136. [CrossRef]
5. Sun, H.; Pi, B.; Sun, J.; Miyamae, T.; Morinaga, M. SASLedger: A Secured, Accelerated Scalable Storage Solution for Distributed Ledger Systems. *Future Internet* **2021**, *13*, 310. [CrossRef]
6. Akbar, N.A.; Muneer, A.; ElHakim, N.; Fati, S.M. Distributed Hybrid Double-Spending Attack Prevention Mechanism for Proof-of-Work and Proof-of-Stake Blockchain Consensuses. *Future Internet* **2021**, *13*, 285. [CrossRef]
7. Xu, R.; Nagothu, D.; Chen, Y. EconLedger: A Proof-of-ENF Consensus Based Lightweight Distributed Ledger for IoVT Networks. *Future Internet* **2021**, *13*, 248. [CrossRef]
8. Yiu, N.C. Toward Blockchain-Enabled Supply Chain Anti-Counterfeiting and Traceability. *Future Internet* **2021**, *13*, 86. [CrossRef]

Review

Blockchain Applications in Agribusiness: A Systematic Review

Geneci da Silva Ribeiro Rocha [1], Letícia de Oliveira [2],* and Edson Talamini [2]

[1] Interdisciplinary Center for Studies and Research in Agribusiness—CEPAN, Universidade Federal do Rio Grande do Sul—UFRGS, Rio Grande do Sul 90040-060, Brazil; geneci.6813.srr@gmail.com

[2] Department of Economics and International Relations—DERI, Faculty of Economics, and Interdisciplinary Center for Studies and Research in Agribusiness—CEPAN, Universidade Federal do Rio Grande do Sul—UFRGS, Rio Grande do Sul 90040-060, Brazil; edson.talamini@ufrgs.br

* Correspondence: leticiaoliveira@ufrgs.br

Abstract: Blockchain is a technology that can be applied in different sectors to solve various problems. As a complex system, agribusiness presents many possibilities to take advantage of blockchain technology. The main goal of this paper is to identify the purposes for which blockchain has been applied in the agribusiness sector, for which a PRISMA-based systematic review was carried out. The scientific literature corpus was accessed and selected from Elsevier's Scopus and ISI of Knowledge's Web of Science (WoS) platforms, using the PRISMA protocol procedures. Seventy-one articles were selected for analysis. Blockchain application in agribusiness is a novel topic, with the first publication dating from 2016. The technological development prevails more than blockchain applications since it has been addressed mainly in the Computer Sciences and Engineering. Blockchain applications for agribusiness management of financial, energy, logistical, environmental, agricultural, livestock, and industrial purposes have been reported in the literature. The findings suggest that blockchain brings many benefits when used in agribusiness supply chains. We concluded that the research on blockchain applications in agribusiness is only at an early stage, as many prototypes are being developed and tested in the laboratory. In the near future, blockchain will be increasingly applied across all economic sectors, including agribusiness, promoting greater reliability and agility in information with a reduced cost. Several gaps for future studies were observed, with significant value for science, industry, and society.

Keywords: Internet of Things; information technology; innovation; supply chains; transparency; traceability; safety; food systems; transactions; smart contracts

Citation: Rocha, G.d.S.R.; de Oliveira, L.; Talamini, E. Blockchain Applications in Agribusiness: A Systematic Review. *Future Internet* 2021, 13, 95. https://doi.org/10.3390/fi13040095

Academic Editor: Ahad ZareRavasan

Received: 19 February 2021
Accepted: 5 April 2021
Published: 8 April 2021

Publisher's Note: MDPI stays neutral with regard to jurisdictional claims in published maps and institutional affiliations.

Copyright: © 2021 by the authors. Licensee MDPI, Basel, Switzerland. This article is an open access article distributed under the terms and conditions of the Creative Commons Attribution (CC BY) license (https://creativecommons.org/licenses/by/4.0/).

1. Introduction

Agribusiness is a sector of paramount importance for a country; its potential is the result of a set of several factors, especially investments in technology and research, which can increase productivity. The agribusiness aggregate is composed of several inputs or product supply chains operating in different natural ecosystems. In a supportive context, the supply chains are embedded in an institutional environment formed by financial, R&D, and technical assistance organizations and institutions with a strong influence on their performance [1].

Supply chains involve a set of activities that are gradually structured from the production to manufacturing and marketing of a product or service. A supply chain is understood to be the operations involving everything from the manufacturing of inputs, production on the farm, transformation process (industrialization), distribution, and trading to reach the end consumer [2]. Thus, the agribusiness supply chains of a country may have great prominence all over the world since they are complex, diversified, and dependent on the organization of each link in the productive system and the relation between them. The farmers operate in a strong, participative, and comprehensive system with an increasingly interconnected integration within a wide business and cooperation network. A farm represents an important node, integrated with other nodes such as infrastructure, commerce, finance, technology, labor relations, and the entire public and private institutional apparatus [2,3]. In addition to the number of actors and transactions, their complexity is increasing over time, taking into account the countless amount of country-specific economic, political, environmental, social, cultural, legal, and sanitary norms and conventions [3].

Many resources are changed through the links between actors involved in an agribusiness supply chain. In one direction, there is a flow of products and services, while in the opposite direction, financial resources are exchanged. Today, information is a valuable resource, which flows both forward and backward [4,5]. Information sharing is subject to many benefits and constraints depending on objectives, interests, technology, trust, and control, among other factors [6]. The asymmetric nature of information and misinformation is a constant challenge to be overcome in supply chain management, and consumers are demanding traceable products, transparency, and safety information [7–10].

The fast growth and development of information technology, attributed to a platform of public services aiming to improve the management of distant resources and services, became the key to solving frictions between the demand and supply of products in organizations [10]. The blockchain is a communication network where data are stored and shared in a distributed manner between all links, eliminating any trusted authority centralized in different business models, and where each node can coordinate without a unified data center [11]. The concept was introduced in 2009 by the pseudonymous Satoshi Nakamoto, who created Bitcoin to solve double-spending problems. The nodes in the Bitcoin network incorporate mutually agreed validations on the blockchain, carrying the transactions on the ledger, that is, a kind of cash book determining who has data in that chain [12].

Blockchain is considered a technological innovation that comes from the incorporation of existing technologies and, lately, has received more attention due to its autonomy, anonymity, and data immutability, becoming an emerging subject in science and organizations [11]. It is a distributed database, in which a group of people controls, registers, and shares information, that can be used in different kinds of applications and is interconnected through platforms and hardware all over the world. It has been identified as a technology that has a concept based on a protocol that is inviolable to human action and it is also based on three underlying technologies: peer-to-peer networks (P2P), cryptography, and distributed consistency algorithms [13]. It is also accompanied by a smart contract, which is not a necessary part of blockchain-based systems but provides natural support for transactions carried out using the technology [14].

A blockchain is a chain of information blocks interconnected in the digital environment of the internet, allowing for information on transactions of various kinds to be stored, linked, and recovered, forming a large database [15]. The blockchain networks can be categorized into public, private, and federated blockchains, based on the network management system adopted and the permissions allowed [16]. A private blockchain is a permissioned access platform; a public blockchain is a permissionless open data network in which any user can add data in the form of a transaction, which is an identification data package in the system, and these data can also be checked and copied; a federated blockchain combines the features of both private and public blockchains [16,17].

Blockchain technology is claimed to be a technology of inviolable validation, having decentralization as a safety measure that creates consensus and confidence in direct communication between two parts, without third party intermediation. It is appropriate for situations requiring privacy, identity control, and permissions [17,18]. Thus, a blockchain provides immutability, transparency, and almost instantaneous insurance in the form of information shared between two or more participants in a single transaction, eliminating the need for third parties, creating an immutable record which can be seen by all the relevant parties without being altered.

When addressing the blockchain theme, it is common to think of cryptocurrencies, especially Bitcoin. However, experts in technology and economics are signaling that blockchains may be relevant in technological research and can be used in many areas in addition to Bitcoin [19]. Many organizations have integrated technologies which have been changing the environment and markets, making it easier to develop activities favoring all parts of the process, including the agribusiness segments, resulting in positive contributions to the agri-food supply chain.

The current agribusiness supply chains are ineffective because their integrity can be easily counterfeited. Blockchain applications have the potential for automated control in which the product is traced from the farm to the consumer's table, depending on the 3P's (party, product, and premises) conditions; with this technology, the end consumer can select the product they want to consume, with safety in the product quality. In addition to the product quality differential, the price received by the producer tends to be higher because of the guarantees. In this way, blockchains can improve traceability in agribusiness, adding value not only for the producer but the entire production chain as well. Thus, an infrastructure supported by the use of this technology can help guarantee food safety, as effective tracking reduces losses in the logistic process [7–10,19–23]. Blockchain applications have been reported for other purposes related to agribusiness, such as precision agriculture and environmental monitoring [24], transactions certification [25], smart agriculture and smart contracts [26], consumer–retailer relationships [27,28], sustainability and coordination, performance, and order management [29], development of safer algorithms [30], and fair-trading practices [31].

Although blockchain is a new technology, its uses and potential applications in agribusiness are already being reported. However, the literature lacks studies that investigate the purposes for which blockchain technology has been employed. Therefore, this study is conducted using the following research question: what are the purposes for which blockchain has been applied in agribusiness supply chains? The main goal is to identify the purposes for which the blockchain has been applied in the agribusiness sector.

The article is structured in five sections. The first section addresses the subject, evidencing the research problem and objective. The second section explains the methodological procedures used to achieve the objective. In the third section, the data analysis is presented and discussed; conclusions, limitations, and suggestions for future studies are offered in the last section.

2. Materials and Methods

In this section, the steps and techniques used to achieve the objective of this research are presented. The study is exploratory and quali-quantitative in its nature. A systematic review of the literature was accomplished to solve the research question and reach the objective. The review was based on articles indexed in the Elsevier's Scopus and ISI of Knowledge's Web of Science (WoS) databases searched on 5 July 2019, with no restriction on the date of publication. Relevant and high-impact studies were selected, taking into account the databases chosen.

Following the procedures stated by the PRISMA protocol, a bibliometric review followed by a systematic content analysis of the articles was carried out. Bibliometrics is a quantitative and statistical technique for systematically measuring production indexes that contribute to the topic in the academic environment, characterized by measuring the influence of journals, researchers, and their trends [32]. The PRISMA protocol helps researchers improve the reporting of systematic reviews or meta-analyses, since it is based on an objective question, uses detailed and clear methods, and allows for identifying, selecting, and critically evaluating the most relevant research on the studied subject [33].

The systematic review is presented in the following steps: (i) formulate the research question; (ii) define the inclusion or exclusion criteria; (iii) select and evaluate the quality of the literature included in the study; and (iv) analyze, synthesize, and disseminate the results [34]. This choice was made due to its reliability as well as the methodological rigor necessary to develop it. Likewise, its scope allows a general analysis of the articles' contents, with a clear structuring of the information found [35]. The analysis continues with two phases of operational procedures:

First phase: two search queries were used in the databases (Scopus and WoS) to retrieve a larger number of studies. The following keywords were inserted in the search engine: the first attempt started using the terms "agr*" and "blockchain", or "block chain" and "block-chain", returning 82 articles in the Scopus database and 221 in WoS, totaling 303 studies. In the second attempt, the search was carried out with the keywords "blockchain", and a sub search inserting the term "agr*"; it found 266 studies in Scopus, and 160 in WoS, totaling 426 articles, as shown in Figure 1. The terms were searched under "title", "abstract", and "keywords", selecting "article" as the type of document.

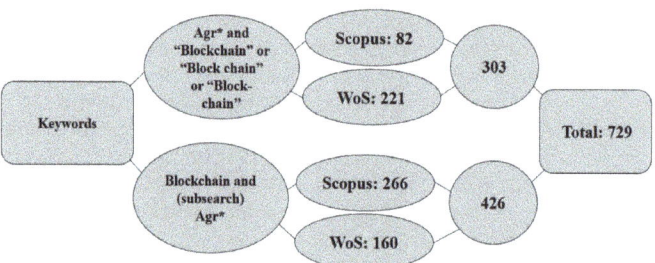

Figure 1. Search criteria and documents retrieved in Scopus and Web of Science databases.

On completing this procedure, 729 articles were found. Then, the methodology of the PRISMA protocol was applied, as shown in Figure 2.

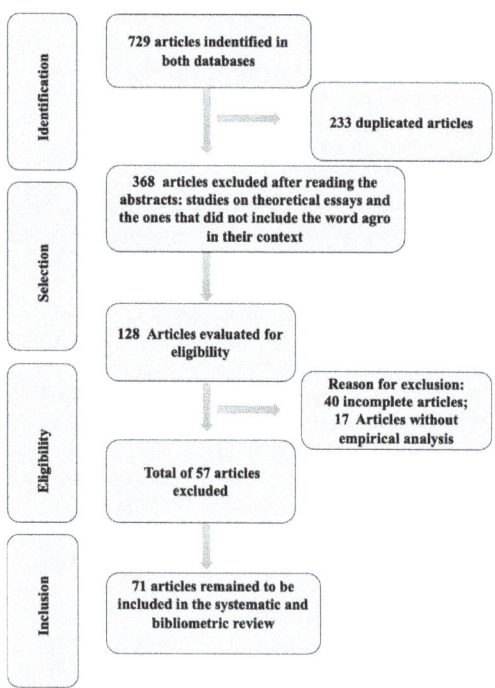

Figure 2. PRISMA flowchart and selection of articles for bibliometric and systematic review.

Figure 2 shows that, of the 729 articles found in the survey, only 71 met the four steps of the protocol: identification, selection, eligibility, and inclusion. The articles that did not have agribusiness as the study objective, those that did not make their full text available on the web, and those dealing with theoretical and bibliometric works, systematic reviews and literature analyses were disregarded. However, the excluded articles contributed to a better comprehension of the subject under study, since they were reviews of blockchain. Thus, during this stage, it was required that the central theme of an article for inclusion was blockchain and agribusiness, being the selected articles on the applications of blockchain for some purpose related to agribusiness.

Second phase: frequency of articles over time, areas of knowledge, number of publications per country, institutions, and authors, as well as the word cloud, were defined as the main bibliometric indicators; objectives, methodological procedures, purposes for blockchain application, and countries where the study was developed were the most important topics for the systematic review. The analysis and discussion of the results were developed through a quali-quantitative analysis, making a systematic review of the articles by reading each of them in depth and extracting fragments of interest.

3. Results and Discussion

In the search for results regarding the purposes for which blockchain has been applied in the agribusiness supply chains, bibliometric analyses and discussions based on a systematic review of the literature were used to find evidence from studies. Therefore, this section presents the main results from the bibliometric analyses, and a qualitative systematic review on the purposes for blockchain application.

3.1. Bibliometric Analysis

The bibliometric results are outlined in different parameters, such as chronological evolution, areas of publication, countries, institutions, and authors that have published on the subject. We found that an article from 2016 was the earliest publication in WoS, when no publication about blockchain in agribusiness was retrieved in the Scopus database. In 2017, two studies in Scopus and five in WoS were identified; in 2018, Scopus presented eight publications and WoS, 13. In 2019, there was a substantial increase to 32 documents in Scopus, and 10 in WoS, showing that the subject is a recent hot topic.

Blockchain in the agribusiness sector is just beginning to be applied since the first article was published in 2016. However, it was observed that most publications appear in 2019 and the Scopus platform stands out in terms of the number of publications. It was deduced that research on the subject has grown significantly in the last three years. This upward trend highlights the emerging nature of blockchain and increasing interest from researchers, universities, and organizations, although it was only introduced in 2009 with the Bitcoin core technology. The academic community, after some years, has identified the potential of blockchain and its possible applications.

Sometimes, blockchain is used as a synonym of Bitcoin, having previously been used only for cryptocurrency applications. In terms of the areas of knowledge publishing on blockchain, it was noted that Engineering (25%), Computer Science (24%), and Social Science (12%) were the most evident in Scopus. With WoS, it is the following: Computer Sciences, 32%; Telecommunications, 17%; Engineering and Environmental Science, 12% (Figure 3). However, knowledge areas vary greatly, enriching the subject interdisciplinarity since some studies are classified in more than one knowledge area. Interdisciplinarity is the meeting of different subjects for the construction of new knowledge, either from a pedagogical or epistemological point of view [36]. This breakdown is justified by the fact that blockchain is a technology that needs software developers to create programs and algorithms that adapt to needs. Likewise, the technology is discussed in relation to operation and development, and the application process is still very restricted. The findings suggest that the subject is still in an exploratory phase where the technology development precedes its application.

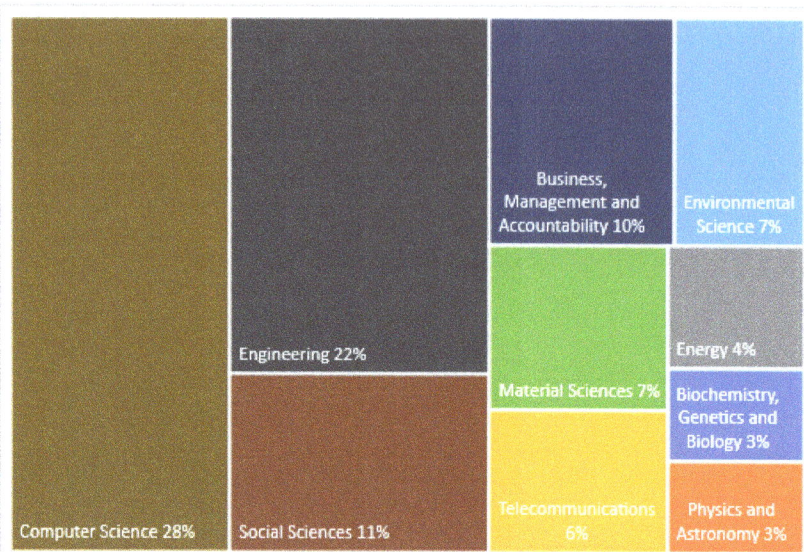

Figure 3. Knowledge areas with the highest number of publications in both databases.

Ten countries stand out, China being the leader in the publishing index, with 39%, and the USA, with 25%. Together, these two countries account for almost two-thirds of publications worldwide. As for the other countries, Australia, South Africa, Finland, the United Kingdom, Italy, and Germany account for 5% each, and Spain and Brazil, account for 3% each. It should be observed that in these countries, there are institutions investing strongly in this technology area, besides being world leaders in technological innovation [37].

The main institutions that stand out in terms of numbers of publications on blockchain in agribusiness are Worcester Polytechnique Institute and Beijing Technology and Business University each with three articles published, and China Agricultural University, National Institute of Industrial Engineering, California State University, Bakersfield, Lancaster University, Shanghai University, Karlsruhe Institute of Technology, Purdue University, and Purdue University System, with two articles each. The first two universities represent the countries that excel in research, located in the USA (Massachusetts) and in China (Pequim), and are known for being prominent in research investment.

The authors that excelled in numbers of publications in Scopus were Hao, Z; Kouhizadeh, M; Mao, D; Wang, F., with three publications, and in WoS, the author Xie, C., with two documents. It should be stressed that only the author Alcarria, R. appeared in both databases, with one publication on each platform, and the others have publications in only one database. Figure 4 shows the last bibliometric indicator of this study, which aimed at generating a word cloud with the most frequent keywords mentioned in the 71 articles.

The word cloud was generated using the 692 keywords extracted from the 71 articles evaluated. It was possible to identify that the most frequent expressions are: Blockchain, Technology, Food, Energy, Smart, Supply Chain, Management, Traceability, System, and Agricultural. It is observed that these words are related to agribusiness, which corroborates the objective of this study. The following is the systematic analysis of the 71 selected studies.

Figure 4. Word cloud of the most frequent keywords in the 71 articles on blockchain in agribusiness.

3.2. A Qualitative Systematic Review

Blockchain applications are pointed out in several sectors, including finance, in which a significant percentage of networks were formed and used Blockchain 1.0 for cryptocurrency [37,38]. Other authors are categorized according to versions "2.0 and 3.0", which are smart contracts encompassing all the financial and economic areas with efficient applications in economics, markets, general sciences, and governmental areas [34–36]. A scheme of the purposes for which blockchain has been applied in agribusiness was constructed in the analysis of the selected articles, and is outlined in Figure 5. The purposes are related to financial, energy, logistical, environmental, agricultural, cattle breeding and livestock, and industrial management, with some specific uses within each segment. It is important to stress that blockchain implementation has been identified in some supply chains, but they are prototypes still in the laboratory tests and applications phase, which are related and exemplified in the next subsection.

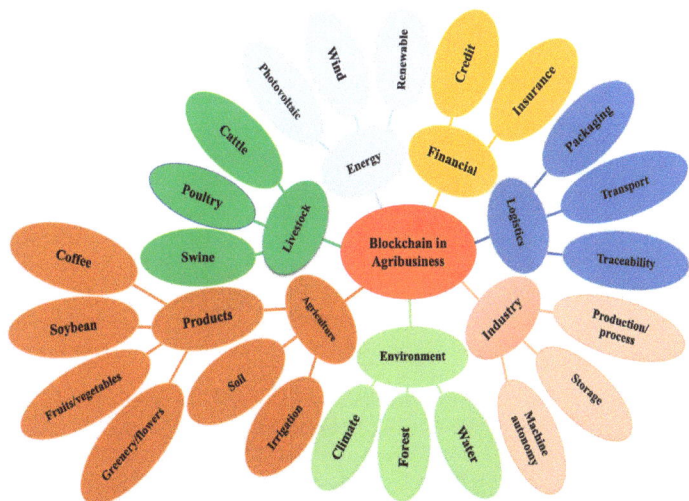

Figure 5. The purposes for which blockchain has been applied in the agribusiness sector.

3.2.1. Blockchain Application for Finance in Agribusiness

Blockchain is considered to be one of the most promising technologies for secure financial transactions for various economic activities, including agribusiness. In agribusiness in particular, it has the potential to revolutionize support for financial transactions for agriculture and improve the credit system. Blockchain allows the reduction of information asymmetry, establishing a fluent channel for transmitting information, increasing the reliability of transactions, improving efficiency, and reducing the costs of agricultural financing [37,38]. The mechanisms for recording, storing, validating, and protecting data are aimed at solving financing problems between all actors in agricultural supply chains. This includes farmers, development agencies, banks, insurance companies, and other financial institutions [37,38]. The AgriDigital and OlivaCoin platforms, both enabled for blockchain, are cited as examples of faster and safer payment opportunities in rural credit management [7,39]. Tripoli and Schmidhuber [40] state that an integrated payment solution has been launched to increase liquidity in private bonds and global payments using distributed ledger technology (DLT) applied in financing and agricultural insurance through smart contracts.

Hu et al. [41] mention the study applied to delay-tolerant payment that focused on remote rural villages in India, which have a community-run station, such as Nokia Kuha, connected to the public internet through unreliable satellite links. Mao et al. [42] implemented a credit assessment system, adopting blockchain to strengthen the supervision and management of traders in the food supply chain where the entire processing flow and logistics are provided by smart contracts. The smart contracts are programs that combine computer protocols from the user interface to execute the terms of a contract. Since, with blockchain, the whole process becomes more simplified, no longer needing intermediaries involved in asset contracts, it controls the damage of the properties, tangible or intangible, by sharing the access data [43].

3.2.2. Blockchain Application for Energy in Agribusiness

The current stage of agricultural development seeks to face the constant increase in production with the use of new equipment and facilities. This makes the agriculture sector a major consumer of energy. In this context, more efficient management of energy resources in agriculture is a fundamental requirement.

Zheng et al. [44] proposed a study to minimize the consumption of photovoltaic energy in greenhouses by using blockchain. In this system, the index that measures the ratio of photovoltaic consumption is defined to assess the condition of consumption in each greenhouse. As a result, the optimal load operation scheme is always recorded on the blockchain. The authors conclude that the blockchain control strategy can effectively improve local energy consumption. This reduces costs with the purchase and losses of energy, improving the quality of electric grid voltage in the farms.

Another application that is also being tested is the development of an irrigation system that minimizes energy consumption, especially in times of drought. In this field of blockchain application, Enescu et al. [45] proposed a study on the use of photovoltaic energy to power a soil improvement system. According to the authors, this technology can help small farmers to better manage water resources during periods of drought.

Blockchain applications for energy purposes have encouraged users to join the renewable energy system. Smart systems are being tested by several industries related to energy production in agriculture. Many of these blockchain applications are still in the pilot project phase, but they already show considerable results in training energy producers and distributors. The technology also acts as a facilitator in the negotiation, bringing greater flexibility, playing an important role in the storage of data, and guaranteeing security in the transactions and commercialization of energy.

Zhang [46] presents a digital coupon or cryptocurrency that could be introduced into the trade of waste, energy and by-products, such as fertilizers or raw materials, among farmers and entrepreneurs. This system could maximize the use of agricultural waste by encouraging farmers and companies to work together in the production and commercialization of energy.

3.2.3. Blockchain Application for Logistics in Agribusiness

The proper management of logistic processes is essential for every agribusiness supply chain. Blockchain has the potential to facilitate and streamline logistic processes in agribusiness supply chains, and these processes can be reproduced in real time due to complete digitalization and its automation. With the help of smart contracts, financial transfers can be optimized, making them easier [47–49]. For those authors, this generates a guarantee of a direct network between the incoming and outgoing of goods, and payment operations can be carried out with more autonomy and efficiency. The invoices that are currently created on paper and sent by post can become unnecessary with the use of technology.

Blockchain can help lower costs, in addition to adding value to products. The consumer benefits from more transparency and control in the quality of the products. One of the critical points in long supply chains is ensuring food safety. Although the development of traceability mechanisms has brought more security to the consumer, blockchain has elements that can make it a great differentiator in this process. This is one of the most relevant topics in blockchain applications in agribusiness supply chains. Traceability via blockchains allows the consumer to track the food from its origin to retail marketing, where all of the information can be shared securely through a blockchain [50]. According to their preliminary analysis, Lucena et al. [51] concluded that traceability and blockchain-based certification can increase grain exports by 15% in the Brazilian context. Given that the blockchain allows for the identification of what type of fertilizer was used to grow a given crop, for example, this information can be accessed by retailers, auditors, and governments, among others interested in the supply chain [52].

Another element that facilitates logistical processes in agribusiness supply chains is smart contracts. Choi et al. [53] applied a mean-variance model to evaluate the risk sensitivity of decision-makers, examining the transactions and previous data using blockchain records. They concluded that this technology could facilitate the use of information and increase knowledge about the demand for the products, besides being accompanied by smart contracts, and it can be automated to the contracting mechanism, improving the efficiency of the processes. Chang et al. [54] developed an alternative project of a private chain to increase transparency and collaboration in product process distribution. The project notifies status alterations to concerned parties specified in the smart contract registry; these are captured in real time and information changes are achieved through the push of a button. This technology system proposed in the blockchain achieves a better level of efficiency for logistics and operations. The participants of the supply chain can reduce costs associated with the manual operations for the confirmation of tracking, with installations of expensive information systems such as the Electronic Data Interchange (EDI) and the Enterprise Resource Planning (ERP).

It is worth noting that blockchains can be used in logistics, product identification, and contract design. They allow better visualization of transactions between buyers and sellers, without the intervention of intermediaries. The use of blockchain-based applications in supply chains can guarantee security and induce more consistent contract management among interested parties. Blockchains improve the management performance of complex supply chains and enhance the customer's services and transportation systems with new decentralized architectures.

3.2.4. Blockchain Application for Environmental Management in Agribusiness

Agribusiness fights against the image that it negatively impacts the environment [55]. Blockchain can play a key role in the environment due to its data monitoring capability. Research suggests that its adoption can make ecological and sustainable practices in the supply chains easier [56,57]. Lin et al. [24] presented an evaluation tool with social and technical requirements applying blockchains in Information and Communications

Technology (ICT) systems in agriculture, using an interface to visualize the application in its operation, which can promote better development and incorporation of data into local climate, energy use, pesticides, soil quality, production costs, and biodiversity conservation.

Spreng and Spreng [58] proposed studying advances in Information Technology (IT) using social media, identifying the feasible options for an alternative global transnational climate policy. In the research, they included consumers, fossil fuel industries, non-governmental environmental organizations (NGOs), insurance industries, IT companies, representatives of public administrations, and UN agencies. The label accurately shows the reliability of climate impact and the total energy incorporated into final goods and services. Since the monitoring system is transparent, consumers have more trust in the label.

Paiva Sobrinho et al. [59] present a management proposal for the Jundiaí River Basin, based on the adoption of a complementary currency created with the support of blockchains. Technological innovation, such as payment schemes for environmental systems (PES), does not depend on the traditional financial system. Thus, it is immune to economic crises because its creation and management are independent of banks, with the rewards being paid to the makers in the form of cryptocurrency.

Figorilli et al. [60] implemented a blockchain prototype for open-source electronic wood traceability, using Radio-Frequency Identification (RFID) Technology, in which the process is tracked from the standing tree to the end user. In the first step, the tree is identified in the forest and this first stripe is associated with the information in the database: marking date, Global Positioning System (GPS) point of the tree, species, diameter, and height. Other information such as cutting, labels, stacking and production flows are detected by an antenna where the information is again associated with the database. Therefore, labels are produced during production and sale and applied on the final products destined for the end consumer.

It was identified that the utilization of blockchain in the environment is essentially focused on payments for environmental services, forest mapping, traceability, and climate and soil control. However, it has the potential to be explored in several other ways, including problems of climate change and carbon sequestration.

3.2.5. Blockchain Application for Agriculture (Farm) in Agribusiness

Agriculture is one of the most relevant fields in a country, since its production provides food security, nutrition, and health of the population, besides maximizing the economy. In recent years, the agricultural sector has adopted different technologies, such as Internet of Things (IoT) and blockchain, to reach higher yields in production processes [61]. Authors such as Tian [17], Li'na et al. [62], Lin et al. [24], and Mondal et al. [63] have been dedicated to researching the use of blockchain in agriculture, essentially directed to the production traceability.

Li'na et al. [62] proposed an agricultural product based on blockchain and logical architecture in the supply chain, aiming at the involvement and adherence of farmers to technologies to minimize the problem of mutual trust. Those authors presented a food safety traceability system based on blockchain and Electronic Product Code (EPC) Information Services. They developed a prototype to track the product from the farm to the end consumer.

Tao et al. [64] suggested a collaborative tracking system based on blockchain and Electronic Product Code Information Services (EPCIS). The system adopts a smart contract at an innovative business level to solve the issues of disclosing information sensitive to data adulteration and reliability. Kamble et al. [65] analyzed smart contracts and the use of blockchain in fruits and vegetables in India, addressing sustainability issues in human relations between facilitators and Agriculture Supply Chain (ASC) practitioners. The intention was to convince the organizations to adopt blockchain in their supply chains to track production from farms to the final consumer.

Tian [47] and Mondal et al. [63] created blockchain inspired by IoT, that is, an architecture to create a transparent food supply chain. For this purpose, they used blockchain, IoT, and RFID indexing terms. According to the authors, the sensing modality was integrated with the identification for tracking and monitoring the quality of the packages, with the food being digitized at different times along the way and the sensor data being updated, in real time, using blockchain, providing a counterfeit-proof digital history. Therefore, any consumer can check, in public accounting, the information on the food packages.

Dos Santos et al. [66] proposed a prototype for the certification and traceability of food products, aiming at tracing the origins of the raw materials without revealing confidential business information. They used the code within Rinkeby Testnet blockchain, applying it to practical examples, including peanuts, cocoa, and apple juice, which are basic ingredients of recipes, thus, demonstrating the feasibility of using the Ingredient Token (IGR) as a methodological figure of certification of ingredients from the farmer to the final consumer, without exposing the formula of the product mixture.

Patil et al. [67] presented a proposal for agriculture with a smart greenhouse, based on activities managed by blockchains, which works through IoT devices. In the project, the authors exhibited this model to achieve security, a light and decentralized privacy project, optimization of resources, and energy consumption. Munir et al. [68] outlined a project, based on an IoT system, used to monitor gardens, controllable from anywhere using a smartphone. The system uses input data collected in real time by sensors that guide farmers and gardeners with a mobile device displaying a list of suggested plants, according to the climate of a particular region.

Scuderi et al. [69] demonstrated the application of blockchain to citrus fruit production in Italy, specifically that referring to Near Field Communication (NFC), in which the process phases involving orange juice are: production, processing, distribution, and sales, with records in a digital profile. According to the authors, key information about citrus farms is stored in digital profiles including the environment and cultivation, soil, water, area, season, plant quality, growing conditions, planting season, and information on the fertilizers and pesticides used for growing oranges. Then, the product is marketed after a digital contract is signed and stored in blockchain.

Borrero [70] tested a traceability system for the food supply chain using blockchain, helping the agricultural cooperatives improve transparency on the origin and processes incorporated into the products. They developed a proof-of-concept blockchain model (PoC) in the agri-food field for the traceability of the supply chain, proving the origin of berry production, and personalizing the roles of each actor in the supply chain. Blockchain, developed with this PoC concept, is being implemented and tested in a Spanish Agricultural Cooperative that uses a book with permissions (hyperledger), based on a smart contract. The demonstration was based on a previous analysis of the berry chain and the interactions between farmers, cooperatives, their certifiers, suppliers, and supermarkets to allow the digital representation of a large number of berries to be associated with a single digital certification.

Salah et al. [71] suggested a system of sequence diagrams based on blockchain for soybean traceability. They also created an algorithm model for selling soy among the several participants, using smart contracts to track and control all interactions and transactions of the participants involved in the ecosystem of the supply chain.

It is inferred that blockchain is being used in the agricultural sector to improve food security, the production process, processing, and transport, besides being applied in a management system of the supply chain to provide transparency, safety, neutrality, and reliability in all the operations of the supply chains. Blockchain, together with IoT, contributes to the resolution of challenges related to operation safety, quality certification, and product origin.

3.2.6. Blockchain Application for Livestock in Agribusiness

The livestock sector is in constant evolution, requiring effective alternatives for its process and management. Advances in transportation and communication technologies promote the development of world markets and facilitate the establishment of animal production units. In that context, several authors have been discussing the use and possibility of blockchain in that sector.

Liu et al. [72] highlight the application of blockchain in the pig meat supply chain by addressing its influence on information sharing and state that it solves problems of matching supply and demand in the supply chain. They added that the transaction costs between the actors involved can be reduced, bringing several benefits: time, money, and improvement for the whole chain through efficient management of the farms.

Sittón-Candanedo et al. [73] implemented a blockchain platform on a dairy farm to minimize costs, through which it was possible to identify and use the available resources more efficiently, besides making it possible to track the animals' locations and monitor their health conditions in real time.

Sander et al. [74] designed a study on the acceptance of blockchain in the traceability and transparency of the meat supply chain. They highlighted the different perspectives and opinions of stakeholders, investigating the transparency and traceability system (TTS) of certification and customers' perceptions as to the potential of technology in traceability. The results of adopting the technology as a solution for the current problems affecting the meat production chain in several countries, including Brazil, are promising.

3.2.7. Blockchain Application for Agroindustry

Blockchain can be used in the food industry in different ways since production goes through processing and final packaging to the consumer. Several pieces of research have been carried out focusing on this sector using, for example, two-dimensional bar code (QR Code) technology, which can already be used in many product packages, since it makes it possible to save a variety of data in a minimal amount of space. The final consumer can view the whole history of the product, from production, processing, to final packaging, on a smartphone [64,65].

The first applications are already being tested in retail companies. Walmart, in partnership with IBM, is using the technology in the traceability of mangoes and pig meat, since it is possible to track the entire supply chain in a few seconds. Furthermore, with blockchain, there is the possibility of identifying contaminated products and eliminating them quickly and efficiently [75]. Blockchain has been also tested for traceability in the milk industry. Behnke and Janssen [76] researched four different dairy industries with distinct features. They investigated every process in the supply chain, from mass production to small product batches, and identified the conditions of information ensuring improvements in batch traceability.

Kouhizadeh et al. [56] propose the application of blockchain in a research context in the circular economy. The authors identified the various ways in which those areas interact between management research and its implications, improving the country's economy and the environment. They conducted the study with initiatives of circular economy at three levels: macro, meso, and micro. Thus, they identified how technology can contribute to the product, its suppression, and its synergies.

Zhu and Kouhizadeh [11] conducted a case study with industries to track the sales performance of their products using blockchain. The uses of blockchain aimed at supplying the following sustainability efforts in the chain: management, product reuse (recycling), regeneration, manufacture, and waste management, which can be tracked for decision making and product disposal. The blockchain can track information from stakeholders such as retailers and customers.

Tallyn et al. [77] developed a Bitbarista prototype with the objective of providing opportunities for the creation of autonomous systems aiming at contributing to the most independent and transparent organizations. The research presented a study conducted in three environments in an office, for one month, exploring the impact of an autonomous coffee machine in the daily activity of coffee consumption. The Bitbarista measures the coffee consumption using autonomous processes in blockchain, presenting the data of origin at the time of the purchase, while aiming to reduce intermediaries in the trade of this product. The report of interactions with the Bitbarista and around it explores the implications for daily life, its structures, and social values.

It has been identified that the technology can be successfully implemented in various sectors and departments of an industry, which allows understanding for how to apply blockchain in the development of daily activities within an industry.

4. Conclusions

Based on the bibliometric analyses, as in the results presented, it was found that the blockchain subject has grown in academia in recent years. However, regarding the analysis of the studies, it was possible to ascertain that, although blockchain emerged more than ten years ago, the discussion of its application in the context of agribusiness is still very recent. Moreover, as the technology development moves first in science, it is possible to say that practical applications in agribusiness supply chains are less advanced.

It was verified that most publications are concentrated in the areas of Computer Science and Engineering, which is justified by the fact that blockchain is a technology that requires software developers to create models adapted to certain demands. Blockchain development is concentrated in countries such as the United States and China. Although American agribusiness is well developed, other countries where the sector plays an important role, such as Brazil, have made a modest contribution to the scientific development of blockchain technology.

It is relevant to note that research on the application of blockchains in agribusiness segments is still at an early stage, including its experimentation. In the case of this study, laboratory prototypes were found in the analysis of the articles, which are in the application testing phase. There are several applications of the technology in the "proof of concept" phase, and few are implemented on a large scale. There are other application possibilities, since its potential at the moment may not be correctly estimated, as it is still an emerging technology.

Financial, energy, logistical, environmental, agricultural, livestock, and industrial are the main purposes for which blockchain has been applied in agribusiness supply chains, as reported in the scientific literature. The purposes are not independent or mutually exclusive of each other. On the contrary, blockchain application in agribusiness supply chains can provide cross-factor benefits. Blockchain applications with a logistical purpose, for instance, may also result in energy, financial, and environmental advantages.

Blockchain is a technology in development within different sectors, with several application possibilities and vast advantages, such as an increase in confidence, reduction of risks in transactions, less bureaucracy, reduction of costs due to the elimination of intermediaries, reduction of risks and frauds, and greater privacy due to the rigorous controls made much more secure due to the immutability of the data. Evidence has been identified that blockchain brings countless benefits when used in agribusiness supply chains by reducing intermediaries and transactional cost, and improving the processes as a whole. It enables the transformation of different transactional processes, making them simpler and faster. Accordingly, the use of blockchain in agribusiness supply chains can facilitate collaboration safely and reliably among the participants involved in business networks. Traceability and smart contracts are frequently reported as advantageous applications in agribusiness.

For future studies, it is suggested that issues, such as transaction costs, information governance, new business models, information asymmetry, and the use of blockchain as a management tool in the agribusiness sectors, be explored. Another area for investigation is the implementation cost of that technology.

Author Contributions: Conceptualization, G.d.S.R.R. and L.d.O.; methodology, G.d.S.R.R. and L.d.O.; software, G.d.S.R.R. and L.d.O.; validation, G.d.S.R.R., L.d.O., and E.T.; formal analysis, G.d.S.R.R., L.d.O., and E.T.; investigation, G.d.S.R.R. and L.d.O.; resources, G.d.S.R.R., L.d.O., and E.T.; data curation, G.d.S.R.R. and L.d.O.; writing—original draft preparation, G.d.S.R.R. and L.d.O.; writing—review and editing, G.d.S.R.R., L.d.O., and E.T.; visualization, G.d.S.R.R. and L.d.O.; supervision, L.d.O.; project administration, G.d.S.R.R. and L.d.O.; funding acquisition, G.d.S.R.R. and E.T. All authors have read and agreed to the published version of the manuscript.

Funding: This study was partially funded by the Coordenação de Aperfeiçoamento de Pessoal de Nível Superior—Brasil (CAPES) through a scholarship granted for the first author (Process Number 88882.439355/2019-01), and by the National Council for Scientific and Technological Development—CNPq through a research grant (Process Number 303956/2019-4).

Institutional Review Board Statement: Not applicable.

Informed Consent Statement: Not applicable.

Data Availability Statement: Not applicable.

Acknowledgments: The author are grateful to the Graduate Program in Agribusiness at the Universidade Federal do Rio Grande do Sul—UFRGS, the Coordenação de Aperfeiçoamento de Pessoal de Nível Superior—Brasil (CAPES), and the National Council for Scientific and Technological Development (CNPq). The authors thank the anonymous reviewers for their comments and suggestions.

Conflicts of Interest: The authors declare no conflict of interest.

References

1. Jank, M.S.; Nassar, A.M.; Tachinardi, M.H. Agronegócio e comércio exterior brasileiro. *Rev. USP* **2005**, *14*. [CrossRef]
2. Furlanetto, E.L.; Cândido, G.A. Metodologia para estruturação de cadeias de suprimentos no agronegócio: Um estudo exploratório. *Rev. Bras. Eng. Agrícola Ambient.* **2006**, *10*, 772–777. [CrossRef]
3. Durski, G.R. Avaliação do desempenho em cadeias de suprimentos. *Rev. FAE* **2003**, *6*, 27–38.
4. Pedroso, M.C.; Nakano, D. Knowledge and information flows in supply chains: A study on pharmaceutical companies. *Int. J. Prod. Econ.* **2009**, *122*, 376–384. [CrossRef]
5. Kaipia, R. Coordinating material and information flows with supply chain planning. *Int. J. Logist. Manag.* **2009**, *20*, 144–162. [CrossRef]
6. Kembro, J.; Näslund, D.; Olhager, J. Information sharing across multiple supply chain tiers: A Delphi study on antecedents. *Int. J. Prod. Econ.* **2017**, *193*, 77–86. [CrossRef]
7. Kamilaris, A.; Fonts, A.; Prenafeta-Boldú, F.X. The rise of blockchain technology in agriculture and food supply chains. *Trends Food Sci. Technol.* **2019**, *91*, 640–652. [CrossRef]
8. Flores, L.; Sanchez, Y.; Ramos, E.; Sotelo, F.; Hamoud, N. Blockchain in Agribusiness Supply Chain Management: A Traceability Perspective. In *Advances in Artificial Intelligence, Software and Systems Engineering*; Ahram, T., Ed.; Springer: Berlin/Heidelberg, Germany, 2021; pp. 465–472.
9. Rejeb, A.; Keogh, J.G.; Treiblmaier, H. Leveraging the Internet of Things and Blockchain Technology in Supply Chain Management. *Future Internet* **2019**, *11*, 161. [CrossRef]
10. Leng, K.; Bi, Y.; Jing, L.; Fu, H.-C.; Van Nieuwenhuyse, I. Research on agricultural supply chain system with double chain architecture based on blockchain technology. *Future Gener. Comput. Syst.* **2018**, *86*, 641–649. [CrossRef]
11. Zhu, Q.; Kouhizadeh, M. Blockchain Technology, Supply Chain Information, and Strategic Product Deletion Management. *IEEE Eng. Manag. Rev.* **2019**, *47*, 36–44. [CrossRef]
12. Christidis, K.; Devetsikiotis, M. Blockchains and Smart Contracts for the Internet of Things. *IEEE Access* **2016**, *4*, 2292–2303. [CrossRef]
13. Laurence, T. *Blockchain for Dummies*; John Wiley & Sons, Inc.: Hoboken, NJ, USA, 2017.
14. Sikorski, J.J.; Haughton, J.; Kraft, M. Blockchain technology in the chemical industry: Machine-to-machine electricity market. *Appl. Energy* **2017**, *195*, 234–246. [CrossRef]
15. Fernández Herrero, D. Aplicación de la Tecnología Blockchain en el Supply Chain en los Sectores Industriales, Escuela de Ingenierías Industriales. Master's Thesis, Univesidad de Valladolid, Valladolid, Spain, 2018.
16. Yang, R.; Wakefield, R.; Lyu, S.; Jayasuriya, S.; Han, F.; Yi, X.; Yang, X.; Amarasinghe, G.; Chen, S. Public and private blockchain in construction business process and information integration. *Autom. Constr.* **2020**, *118*, 103276. [CrossRef]
17. Tian, F. An information System for Food Safety Monitoring in Supply Chains Based on HACCP, Blockchain and Internet of Things. Ph.D. Thesis, WU Vienna University of Economics and Business, Vienna, Austria, 2018.
18. Atzori, M. Blockchain technology and decentralized governance: Is the state still necessary? *J. Gov. Regul.* **2017**, *6*, 45–62. [CrossRef]
19. Casino, F.; Dasaklis, T.K.; Patsakis, C. A systematic literature review of blockchain-based applications: Current status, classification and open issues. *Telemat. Inform.* **2019**, *36*, 55–81. [CrossRef]
20. Ferreira, J.E.; Pinto, F.G.C.; dos Santos, S.C. Estudo de Mapeamento Sistemático sobre as Tendências e Desafios do Blockchain. *Rev. Gestão Org.* **2017**, *15*, 108–117. [CrossRef]
21. Lezoche, M.; Hernandez, J.E.; Díaz, M.M.E.A.; Panetto, H.; Kacprzyk, J. Agri-food 4.0: A survey of the supply chains and technologies for the future agriculture. *Comput. Ind.* **2020**, *117*, 103187. [CrossRef]
22. Xie, C.; Sun, Y.; Luo, H. Secured Data Storage Scheme Based on Block Chain for Agricultural Products Tracking. In Proceedings of the 2017 3rd International Conference on Big Data Computing and Communications (BIGCOM), Chengdu, China, 10–11 August 2017; IEEE: Piscataway, NJ, USA, 2017; pp. 45–50.

23. Papa, S.F. Use of Blockchain Technology in Agribusiness: Transparency and Monitoring in Agricultural Trade. In Proceedings of the 2017 International Conference on Management Science and Management Innovation (MSMI 2017), Suzhou, China, 23–25 June 2017; Atlantis Press: Paris, France, 2017.
24. Lin, Y.-P.; Petway, J.; Anthony, J.; Mukhtar, H.; Liao, S.-W.; Chou, C.-F.; Ho, Y.-F. Blockchain: The Evolutionary Next Step for ICT E-Agriculture. *Environments* **2017**, *4*, 50. [CrossRef]
25. Tse, D.; Zhang, B.; Yang, Y.; Cheng, C.; Mu, H. Blockchain application in food supply information security. In Proceedings of the 2017 IEEE International Conference on Industrial Engineering and Engineering Management (IEEM), Singapore, 10–13 December 2017; IEEE: Piscataway, NJ, USA, 2017; pp. 1357–1361.
26. Xiong, H.; Dalhaus, T.; Wang, P.; Huang, J. Blockchain Technology for Agriculture: Applications and Rationale. *Front. Blockchain* **2020**, *3*. [CrossRef]
27. Shew, A.M.; Snell, H.A.; Nayga, R.M.; Lacity, M.C. Consumer valuation of blockchain traceability for beef in the United States. *Appl. Econ. Perspect. Policy* **2021**, aepp.13157. [CrossRef]
28. Garaus, M.; Treiblmaier, H. The influence of blockchain-based food traceability on retailer choice: The mediating role of trust. *Food Control* **2021**, 108082. [CrossRef]
29. Lim, M.K.; Li, Y.; Wang, C.; Tseng, M.-L. A literature review of blockchain technology applications in supply chains: A comprehensive analysis of themes, methodologies and industries. *Comput. Ind. Eng.* **2021**, *154*, 107133. [CrossRef]
30. Fu, X.; Wang, H.; Shi, P. A survey of Blockchain consensus algorithms: Mechanism, design and applications. *Sci. China Inf. Sci.* **2021**, *64*, 121101. [CrossRef]
31. Kononets, Y.; Treiblmaier, H. The potential of bio certification to strengthen the market position of food producers. *Mod. Supply Chain Res. Appl.* **2020**. [CrossRef]
32. Araújo, C.A. Bibliometria: Evolução histórica e questões atuais. *Em Questão* **2006**, *12*, 11–32.
33. Moher, D.; Liberati, A.; Tetzlaff, J.; Altman, D.G. Preferred Reporting Items for Systematic Reviews and Meta-Analyses: The PRISMA Statement. *PLoS Med.* **2009**, *6*, e1000097. [CrossRef]
34. Cronin, P.; Ryan, F.; Coughlan, M. Undertaking a literature review: A step-by-step approach. *Br. J. Nurs.* **2008**, *17*, 38–43. [CrossRef]
35. Tranfield, D.; Denyer, D.; Smart, P. Towards a Methodology for Developing Evidence-Informed Management Knowledge by Means of Systematic Review. *Br. J. Manag.* **2003**, *14*, 207–222. [CrossRef]
36. Bispo, E.P.d.F.; Tavares, C.H.F.; Tomaz, J.M.T. Interdisciplinaridade no ensino em saúde: O olhar do preceptor na Saúde da Família. *Interface Commun. Saúde Educ.* **2014**, *18*, 337–350. [CrossRef]
37. Tapscott, D.; Tapscott, A. How Blockchain Will Change Organizations. *MIT Sloan Manag. Rev.* **2017**, *58*, 10–13.
38. Swan, M. *Blockchain—Blueprint for a New Economy*; O'Reilly Media, Inc.: Sebastopol, CA, USA, 2015.
39. Haferkorn, M.; Quintana Diaz, J.M. Seasonality and Interconnectivity Within Cryptocurrencies—An Analysis on the Basis of Bitcoin, Litecoin and Namecoin. In *Enterprise Applications and Services in the Finance Industry*; Lugmayr, A., Ed.; Springer: Berlin/Heidelberg, Germany, 2015; pp. 106–120.
40. Tripoli, M.; Schmidhuber, J. *Emerging Opportunities for the Application of Blockchain in the Agri-Food Industry*; FAO: Rome, Italy, 2018; 40p.
41. Hu, Y.; Manzoor, A.; Ekparinya, P.; Liyanage, M.; Thilakarathna, K.; Jourjon, G.; Seneviratne, A. A Delay-Tolerant Payment Scheme Based on the Ethereum Blockchain. *IEEE Access* **2019**, *7*, 33159–33172. [CrossRef]
42. Mao, D.; Wang, F.; Hao, Z.; Li, H. Credit Evaluation System Based on Blockchain for Multiple Stakeholders in the Food Supply Chain. *Int. J. Environ. Res. Public Health* **2018**, *15*, 1627. [CrossRef] [PubMed]
43. Nofer, M.; Gomber, P.; Hinz, O.; Schiereck, D. Blockchain. *Bus. Inf. Syst. Eng.* **2017**, *59*, 183–187. [CrossRef]
44. Zheng, C.; Jianhua, Y.; Kaiyuan, J.; Bin, H.; Weizhou, W. Control strategy of time-shift facility agriculture load and photovoltaic local consumption based on energy blockchain. *Electr. Power Autom. Equip.* **2021**, *41*, 47–55. [CrossRef]
45. Enescu, F.M.; Bizon, N.; Stirbu, C. Smart Energy Grids used in irrigation systems using the blockchain applications. In Proceedings of the 2019 11th International Conference on Electronics, Computers and Artificial Intelligence (ECAI), Pitesti, Romania, 27–29 June 2019; IEEE: Piscataway, NJ, USA, 2019; pp. 1–6.
46. Zhang, D. Application of Blockchain Technology in Incentivizing Efficient Use of Rural Wastes: A case study on Yitong System. *Energy Procedia* **2019**, *158*, 6707–6714. [CrossRef]
47. Tian, F. An agri-food supply chain traceability system for China based on RFID & blockchain technology. In *Proceedings of the 2016 13th International Conference on Service Systems and Service Management (ICSSSM), Kunming, China, 24–26 June 2016*; IEEE: Piscataway, NJ, USA, 2016; pp. 1–6.
48. Raskin, M. The law and legality of Smart Contracts. *Georg. Law Technol. Rev.* **2017**, *1*, 305–341.
49. Mao, D.; Hao, Z.; Wang, F.; Li, H. Novel Automatic Food Trading System Using Consortium Blockchain. *Arab. J. Sci. Eng.* **2019**, *44*, 3439–3455. [CrossRef]
50. Bechtsis, D.; Tsolakis, N.; Bizakis, A.; Vlachos, D. A Blockchain Framework for Containerized Food Supply Chains. *Comput. Aided Chem. Eng.* **2019**, *46*, 1369–1374. [CrossRef]
51. Lucena, P.; Binotto, A.P.D.; da Silva Momo, F.; Kim, H. A Case Study for Grain Quality Assurance Tracking based on a Blockchain Business Network. *arXiv* **2018**, arXiv:1803.07877.

52. Janssen, S.J.C.; Porter, C.H.; Moore, A.D.; Athanasiadis, I.N.; Foster, I.; Jones, J.W.; Antle, J.M. Towards a new generation of agricultural system data, models and knowledge products: Information and communication technology. *Agric. Syst.* **2017**, *155*, 200–212. [CrossRef]
53. Choi, T.-M.; Wen, X.; Sun, X.; Chung, S.-H. The mean-variance approach for global supply chain risk analysis with air logistics in the blockchain technology era. *Transp. Res. Part E Logist. Transp. Rev.* **2019**, *127*, 178–191. [CrossRef]
54. Chang, S.E.; Chen, Y.-C.; Lu, M.-F. Supply chain re-engineering using blockchain technology: A case of smart contract based tracking process. *Technol. Forecast. Soc. Chang.* **2019**, *144*, 1–11. [CrossRef]
55. Rajão, R.; Soares-Filho, B.; Nunes, F.; Börner, J.; Machado, L.; Assis, D.; Oliveira, A.; Pinto, L.; Ribeiro, V.; Rausch, L.; et al. The rotten apples of Brazil's agribusiness. *Science* **2020**, *369*, 246–248. [CrossRef]
56. Kouhizadeh, M.; Sarkis, J.; Zhu, Q. At the Nexus of Blockchain Technology, the Circular Economy, and Product Deletion. *Appl. Sci.* **2019**, *9*, 1712. [CrossRef]
57. Saberi, S.; Kouhizadeh, M.; Sarkis, J.; Shen, L. Blockchain technology and its relationships to sustainable supply chain management. *Int. J. Prod. Res.* **2019**, *57*, 2117–2135. [CrossRef]
58. Spreng, C.P.; Spreng, D. Paris is not enough: Toward an Information Technology (IT) enabled transnational climate policy. *Energy Res. Soc. Sci.* **2019**, *50*, 66–72. [CrossRef]
59. Paiva Sobrinho, R.; Garcia, J.R.; Maia, A.G.; Romeiro, A.R. Tecnologia Blockchain: Inovação em Pagamentos por Serviços Ambientais. *Estud. Avançados* **2019**, *33*, 151–176. [CrossRef]
60. Figorilli, S.; Antonucci, F.; Costa, C.; Pallottino, F.; Raso, L.; Castiglione, M.; Pinci, E.; Del Vecchio, D.; Colle, G.; Proto, A.; et al. A Blockchain Implementation Prototype for the Electronic Open Source Traceability of Wood along the Whole Supply Chain. *Sensors* **2018**, *18*, 3133. [CrossRef] [PubMed]
61. Pivoto, D.; Waquil, P.D.; Talamini, E.; Finocchio, C.P.S.; Dalla Corte, V.F.; de Vargas Mores, G. Scientific development of smart farming technologies and their application in Brazil. *Inf. Process. Agric.* **2018**, *5*, 21–32. [CrossRef]
62. Li'na, Y.; Guofeng, Z.; Jingdun, J.; Wanlin, G.; Ganghong, Z.; Sha, T. Modern Agricultural Product Supply Chain Based on Block Chain Technology. *Trans. Chin. Soc. Agric. Mach.* **2017**, *48*, 387–393. [CrossRef]
63. Mondal, S.; Wijewardena, K.P.; Karuppuswami, S.; Kriti, N.; Kumar, D.; Chahal, P. Blockchain Inspired RFID-Based Information Architecture for Food Supply Chain. *IEEE Internet Things J.* **2019**, *6*, 5803–5813. [CrossRef]
64. Tao, Q.; Cui, X.; Huang, X.; Leigh, A.M.; Gu, H. Food Safety Supervision System Based on Hierarchical Multi-Domain Blockchain Network. *IEEE Access* **2019**, *7*, 51817–51826. [CrossRef]
65. Kamble, S.S.; Gunasekaran, A.; Sharma, R. Modeling the blockchain enabled traceability in agriculture supply chain. *Int. J. Inf. Manag.* **2020**, *52*, 101967. [CrossRef]
66. Dos Santos, R.; Torrisi, N.; Yamada, E.; Pantoni, R. IGR Token-Raw Material and Ingredient Certification of Recipe Based Foods Using Smart Contracts. *Informatics* **2019**, *6*, 11. [CrossRef]
67. Patil, A.S.; Tama, B.A.; Park, Y.; Rhee, K.-H. A Framework for Blockchain Based Secure Smart Green House Farming. In *Advances in Computer Science and Ubiquitous Computing*; Park, J., Loia, V., Yi, G., Sung, Y., Eds.; Springer: Berlin/Heidelberg, Germany, 2018; pp. 1162–1167.
68. Munir, M.S.; Bajwa, I.S.; Cheema, S.M. An intelligent and secure smart watering system using fuzzy logic and blockchain. *Comput. Electr. Eng.* **2019**, *77*, 109–119. [CrossRef]
69. Scuderi, A.; Foti, V.; Timpanaro, G. The supply chain value of POD and PGI food products through the application of blockchain. *Qual. Success* **2019**, *20*, 580–587.
70. Borrero, J.D. Sistema de trazabilidad de la cadena de suministro agroalimentario para cooperativas de frutas y hortalizas basado en la tecnología Blockchain. *CIRIEC-España Rev. Econ. Pública Soc. Coop.* **2019**, *71*. [CrossRef]
71. Salah, K.; Nizamuddin, N.; Jayaraman, R.; Omar, M. Blockchain-Based Soybean Traceability in Agricultural Supply Chain. *IEEE Access* **2019**, *7*, 73295–73305. [CrossRef]
72. Liu, L.; Li, F.; Qi, E. Research on Risk Avoidance and Coordination of Supply Chain Subject Based on Blockchain Technology. *Sustainability* **2019**, *11*, 2182. [CrossRef]
73. Sittón-Candanedo, I.; Alonso, R.S.; Corchado, J.M.; Rodríguez-González, S.; Casado-Vara, R. A review of edge computing reference architectures and a new global edge proposal. *Future Gener. Comput. Syst.* **2019**, *99*, 278–294. [CrossRef]
74. Sander, F.; Semeijn, J.; Mahr, D. The acceptance of blockchain technology in meat traceability and transparency. *Br. Food J.* **2018**, *120*, 2066–2079. [CrossRef]
75. Yiannas, F. A New Era of Food Transparency Powered by Blockchain. *Innov. Technol. Gov. Glob.* **2018**, *12*, 46–56. [CrossRef]
76. Behnke, K.; Janssen, M.F.W.H.A. Boundary conditions for traceability in food supply chains using blockchain technology. *Int. J. Inf. Manag.* **2020**, *52*, 101969. [CrossRef]
77. Tallyn, E.; Pschetz, L.; Gianni, R.; Speed, C.; Elsden, C. Exploring Machine Autonomy and Provenance Data in Coffee Consumption. *Proc. ACM Hum.-Comput. Interact.* **2018**, *2*, 1–25. [CrossRef]

Review

Distributed Ledger Technology Review and Decentralized Applications Development Guidelines

Claudia Antal, Tudor Cioara *, Ionut Anghel, Marcel Antal and Ioan Salomie

Computer Science Department, Technical University of Cluj-Napoca, Memorandumului 28, 400114 Cluj-Napoca, Romania; claudia.pop@cs.utcluj.ro (C.A.); ionut.anghel@cs.utcluj.ro (I.A.); marcel.antal@cs.utcluj.ro (M.A.); ioan.salomie@cs.utcluj.ro (I.S.)
* Correspondence: tudor.cioara@cs.utcluj.ro; Tel.: +40-264-202-352

Citation: Antal, C.; Cioara, T.; Anghel, I.; Antal, M.; Salomie, I. Distributed Ledger Technology Review and Decentralized Applications Development Guidelines. *Future Internet* **2021**, *13*, 62. https://doi.org/10.3390/fi13030062

Academic Editors: Sk. Md. Mizanur Rahman and Ahad ZareRavasan

Received: 26 January 2021
Accepted: 24 February 2021
Published: 27 February 2021

Publisher's Note: MDPI stays neutral with regard to jurisdictional claims in published maps and institutional affiliations.

Copyright: © 2021 by the authors. Licensee MDPI, Basel, Switzerland. This article is an open access article distributed under the terms and conditions of the Creative Commons Attribution (CC BY) license (https://creativecommons.org/licenses/by/4.0/).

Abstract: The Distributed Ledger Technology (DLT) provides an infrastructure for developing decentralized applications with no central authority for registering, sharing, and synchronizing transactions on digital assets. In the last years, it has drawn high interest from the academic community, technology developers, and startups mostly by the advent of its most popular type, blockchain technology. In this paper, we provide a comprehensive overview of DLT analyzing the challenges, provided solutions or alternatives, and their usage for developing decentralized applications. We define a three-tier based architecture for DLT applications to systematically classify the technology solutions described in over 100 papers and startup initiatives. Protocol and Network Tier contains solutions for digital assets registration, transactions, data structure, and privacy and business rules implementation and the creation of peer-to-peer networks, ledger replication, and consensus-based state validation. Scalability and Interoperability Tier solutions address the scalability and interoperability issues with a focus on blockchain technology, where they manifest most often, slowing down its large-scale adoption. The paper closes with a discussion on challenges and opportunities for developing decentralized applications by providing a multi-step guideline for decentralizing the design and implementation of traditional systems.

Keywords: distributed ledger technology; blockchain; decentralized applications; technology review; development guidelines; architecture

1. Introduction

Distributed Ledger Technology (DLT) is a disruptive technology that provides an environment with no central authority for registering, sharing, and synchronizing transactions on digital assets. By joining several computer sciences disciplines such as distributed systems, cryptography, data structures, or consensus algorithms, it offers highly desirable features (decentralization, openness, immutability, transparency, traceability, security, availability, etc.). Gartner has included DLT technology in the hype cycle for the first time in 2016 at the phase of an innovation trigger [1] mostly due to the advent of its most popular type the blockchain technology. In 2017–2018, research has been committed to developing the mechanisms to accommodate the technology to requirements such as privacy, scalability, permissions, and interoperability which are essential to decentralized applications implementation perspective and emerging business models. By 2018 the DLT has passed the peak of inflated expectations, assuming to reach a plateau of productivity in the next 5 to 10 years [2], while in 2019 the innovation potential of the DLT is considered by Gartner to be driven not only by the technology expectations but also by the social ones [3].

In this sense, the development of decentralized autonomous organizations and implementation of decentralized applications and the decentralized web is of high interest for the next 10 years. Building such decentralized applications is not a straightforward process since many technological solutions have emerged, generating a confusing context with lots of challenges for the software industry [4].

The main contributions of this paper are a comprehensive overview of nowadays DLT solutions and a set of guidelines for decentralized application development. To streamline and organize the review we have defined and used a three-tier conceptual architecture [5,6] (see Figure 1). Each tier aggregates alternative DLT solutions to address specific issues in the implementation of decentralized applications:

- The Protocol and Network Tier (PN-Tier) aggregates the core DLT elements and organizes them in two layers. The Protocol Layer contains technology solutions for digital assets registration, transactions, data structures, privacy, and business rules implementation. The Network Layer contains technology solutions for creating a peer-to-peer network, ledger replication, and consensus-based validation.
- The Scalability Tier (S-Tier) runs most of the time a parallel DLT network and aggregates technological solutions for addressing the scalability issues raised by the PN-Tier. We have focused on solutions for blockchain ledger scalability problems such as storage scalability, transaction throughput, and computational scalability.
- The Interoperability Tier built on top of the previous two tiers addresses integration and interoperability of multiple DLT applications and systems deployments.

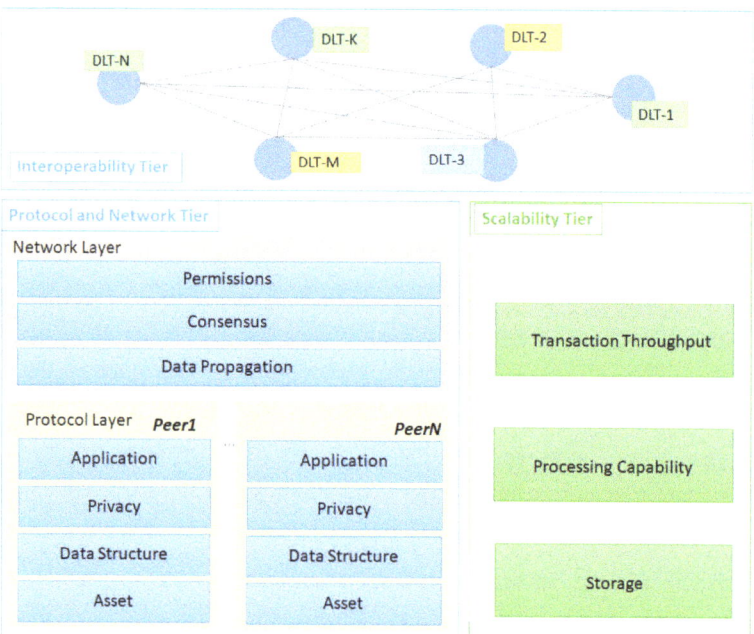

Figure 1. Three-tier architecture for decentralized applications development.

For each architectural tier, we have identified the main technological challenges and used them as criteria of including relevant solutions, and technologies aiming to create a consistent overview of the current technological state of the art. In the case of Scalability and Interoperability Tiers, the focus is mostly on blockchain technology. The reason for this is the higher level of maturity the applications developed using blockchain are often facing problems related to the transaction throughput, storage space, gas consumption, or interoperability with other chains.

Finally, we provide a guideline for building decentralized applications discussing the main steps and the technologies selection concerning the challenges that need to be addressed.

Most distributed ledger technology reviews found in the literature are focusing almost exclusively on the blockchain ledgers [7,8] and do not consider the other DLT variations such as Directed Acyclic Graphs (DAG) [9] and other combinations or hybrid solutions [10]. In our review, we have addressed these DLT variations in the Protocol and Network Tier. Even though in the Scalability and Interoperability Tiers we focus on the blockchain ledgers, being the more mature technology, some of the solutions and patterns referenced can be adopted by other DLT variations. At the same time, our review was driven by a decentralized application architecture that has been used and validated in previous publications. We have successfully used it for the implementation of decentralized applications in the domains of smart energy grid (demand response [6], peer to peer energy trading [11], flexibility management [12], etc.), stock exchange [13], or vaccine distribution [14]. As result, the criteria for selecting the technologies and organizing the review are strictly related to the development issues that may be encountered and need to be addressed in each tier. Finally, the guideline for decentralized application development provided on top of the reviewed literature can drive the selection of various DLT alternatives based on the issue encountered helping the community orienting in such an effervescent and confusing technological context. We could not find a similar guideline in the reviewed literature.

The rest of the paper is structured as follows: Section 2 presents the technological solutions for the PN-Tier, Section 3 reviews the mechanisms of S-Tier for improving the PN-Tier's scalability in three directions: storage, transaction throughput, and computational, Section 4 describes the interoperability solutions among federated blockchain ledgers, Section 5 discusses guidelines for developing decentralized applications, while Section 6 concludes the paper.

2. Protocol and Network Tier

Different solutions for PN-Tier have been proposed addressing specific challenges [4,15–19]. We have identified the technological components that are grouped at the protocol and network layers. The security of the entire tier is ensured as a result of integrating the public-private key cryptography for locking and unlocking transactions, with the tamper-resistant data structures and consensus algorithms.

2.1. Protocol Layer

The protocol layer represents the core of the technology that runs on each full node in the peer-to-peer network. It is governed by rules that specify what, when, how, and by whom the assets are operated on the chain and features four types of technological components: asset representation, data structure, privacy, and business rules enforcement.

2.1.1. Type of Asset and Data Structures

The tokenization process refers to the possibility of modeling different goods in a DLT system as digital assets that can be issued and transferred according to a predefined set of rules. The common terminology for an asset representation in DLT systems is Token or Coin. We have identified two types of tokens that can be represented in the system (see Table 1): native tokens and tokens based on real assets (asset-based tokens). Native tokens are tokens defined in the DLT system, completely independent of the real world, thus the rules governing the issuance and the transfer are completely defined in the system (through Initial Coin Offerings or mining reward schemes), and do not rely on any third trusted party. Bitcoin [4], Ether [15], EtherTulips [20], Grid [21], Rarible [22], CryptoKitties [23], NRGcoin [24] or Telcoin [25] are examples of such tokens that are completely virtual and have economic value based on supply and demand.

Table 1. Type of digital assets modeled using blockchain.

Type	Token Example	Fungibility	Issuers
Native tokens	Bitcoin, Ether, CryptoKitties [4,15,23]	Yes	Mining Reward Schemes
	ERC20, ERC223, ERC-621 [26]	Yes	Initial Coin Offerings
	ERC721 [26]	No	Initial Coin Offerings
Asset-Based Tokens	Real Estate [27]	No	Government Land Registries [28]
	Patents [29]	No	U.S. Patent & Trademark Office [30]
	Academic Records [31]	No	The Registrar's Office [32]
	Gold	Yes	Royal Mint Gold [33]

Real-life assets can also be represented through tokens in DLT systems, offering the opportunity to represent, transfer and track them. By using asset-based tokens, the chain can keep track of different kinds of assets, both tangible (real estate, cars, money, art, etc.) and intangible (patents, trademarks, copyrights, etc.). The DLT systems representing real-life assets must rely on a trusted third party to issue each token concerning the real object. Similarly, a transfer on-chain must be done under the governance of such a trusted party since any issue regarding a wrongful transfer on the chain can be verified only by communicating with the external systems.

In DLT systems, the real-life flow of assets is represented in the network as transactions between peers. The DLT requires specialized data structures at its core that can ensure three properties over the stored transactions: provenance, asset ownership validation, and immutability. Hash pointers have been frequently integrated with different data structures intended for DLT usages. Due to the hash functions' collision-resistant, data concealing, and data binding properties, the hash pointers are the best choice for adapting common data structures to the DLT requirements thus obtaining: linked list with hash pointers (e.g., blockchain), binary trees with hash pointers (e.g., Merkle Trees), graphs with hash pointers (e.g., Hash Graphs), etc.

Two main directions identified are related to distributed ledger data structures for representing peers' transactions, namely blockchain and direct acyclic graph (DAG).

The blockchain structure, as its name suggests, is a chain formed by linked back blocks, also known as Linked List using hash pointers (see Figure 2a). Each block contains all the transactions that occurred in the system in a short period (e.g., ~10 min for Bitcoin, or ~12 s for Ethereum). All the transactions contained in the block are hashed together in a Merkle Tree data structure, where the root of the tree is referenced in the block header and acts as a digital fingerprint of the entire collection. Thus, blockchain becomes an append-only data structure that gathers all the benefits of the hashing and cryptographic functions and, with the integration of a consensus algorithm, ensures an immutable history log of the entire activity of the network. The blockchain structure is implemented in well-known solutions such as Bitcoin [4], Ethereum [15], Litecoin [17], Hyperledger [34], CryptoNode [35], Ripple [36], and Zerocash [19].

A less popular data structure is DAG firstly mentioned in [37]. Since its proposal, several platforms have been developing solutions based on DAG variations: Dagcoin [38], IOTA [16], HashGraph [39], or hybrid systems like Holochain [40], and Flowchain [41]. Among the DAG-based systems, IOTA is the most used solution. It uses a DAG, called tangle, as a ledger for storing the transactions as depicted in Figure 2b. The entire graph starts with a genesis transaction that is approved directly or indirectly by all the transactions in the graph. Whenever a new transaction is submitted, it must validate and confirm two previous transactions from the graph that were not yet approved (i.e., tips). The tips selection algorithm is based on the family of Markov Chain Monte Carlo algorithms and it considers the cumulative weights of sub-tangles. Whenever a situation of conflicting transactions appears, the higher the cumulative weight of the transaction is, the more secure it is.

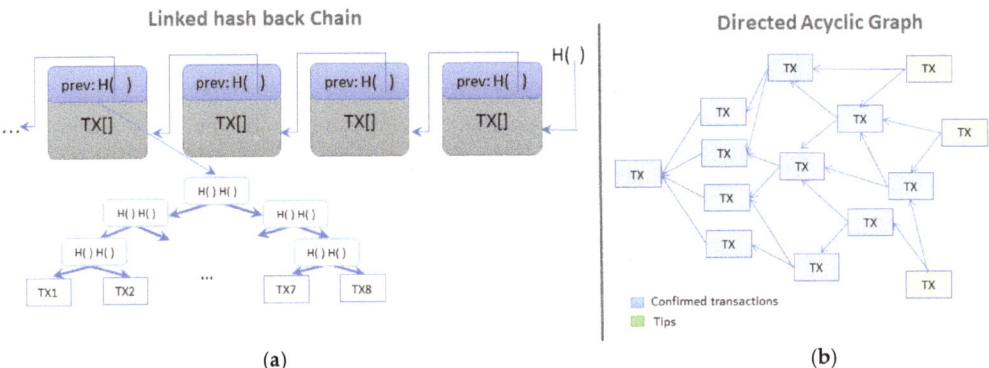

Figure 2. Data for representing peers' transactions structures: (**a**) blockchain using linked lists and (**b**) Directed Acyclic Graphs (DAG).

While the most successful solutions and research directions are focused on blockchain-based DLT, the DAG solutions have yet to overcome considerable shortcomings in terms of centralization and scalability to be considered suitable alternatives to blockchain-based DLTs [42].

2.1.2. Privacy of Transacting Parties and Data

Most of the popular DLT solutions preach the properties of transparency and openness to be major benefits brought by the technology. By providing open information about the transacting parties (pseudo-anonymous most of the time) and transacted assets, the systems offer clear history and audit advantages. However, in many use cases where the privacy of data is highly required [43] (e.g., medical use cases [44,45]), many of the DLT solutions may prove to be unsuitable due to their transparency and openness. With the emerging General Data Protection Regulation (GDPR) restrictions, the privacy of data is one of the most controversial subjects in the DLT systems. In this direction, solutions have emerged that aim to hide the details regarding the transaction information, while at the same time keeping the reliability of the system by allowing consistency checks, validation, and audit regarding the previous actions and history [46].

In Bitcoin versions of the DLT, the value of the transacted asset is clearly stated, and the nodes can easily check the ownership over the asset by checking the history and identifying the unspent transaction outputs (UTXO). The biggest challenge is that by hiding the transaction data, the nodes in the system should still be able to validate the availability of the funds and the ownership of the sender over that asset. Coin Mixers have been developed over Bitcoin to hide information and prevent tracking and tracing of transacted assets. One of these approaches is CoinJoin [47] which aims to aggregate multiple users to agree upon several inputs and outputs and then sign the transactions. This makes it harder for the coins to be traced across the transactions since the group of inputs provided to a transaction will not belong to the actual sender. Similarly, Dash [48] aims to provide mixing services but instead of using a central point that is responsible for mixing inputs and outputs like CoinJoin, it provides a second layer of master nodes over Bitcoin.

A private-public key cryptographic mechanism is proposed by Quorum [49] to hide the transacted value. The encrypted transactions are shared point-to-point only to the involved parties, and a private state is defined by the Quorum node, where these encrypted transactions are stored. The encryption keys are shared between the involved parties, to validate the transaction. However, the shared blockchain (public state) only contains the hash of the encrypted transaction. This leads to a shortcoming of the system since the network cannot verify the validity of the private transactions, this being done only by the involved parties. The shortcoming of Quorum is overcome by integrating Zero-Knowledge proof mechanisms [50] with DLT systems.

Zero-knowledge proofs have been added to digital currencies, such as Zerocoin [19,51] to create anonymous Bitcoin transactions without the need of third parties such as CoinJoin and Dash. The Zero-Knowledge proof mechanisms, mainly succinct Non-Interactive Zero-Knowledge proofs (ZkSnarks) [52] are used to ensure the transactions' validation without revealing actual information about the parties involved. The Zero-Knowledge proofs require that a Verifier could easily check that the Prover owns a secret, without revealing the actual secret to the Verifier. Consider a simple scenario, where the network owns a hash value H. An actor wants to prove to the network that he holds the secret s, that hashed offers the value of H. Normally, the actor would need to reveal the secret to the network. ZkSnarks aim to enhance this model, by allowing the actor to prove ownership over the data without disclosing any information regarding the secret. In this sense, ZkSnarks introduces a proving function, that can be used by the actor to issue a proof showing that the private secret is indeed corresponding to the public information H and a verification function that can be executed by any participant in the network to validate whether the proof corresponds to the public information H. ZCash [19] implements a ZkSnarks mechanism as an improvement of the Bitcoin system that ensures the privacy of the transactions. It requires any transaction to be locked by a secret, called a commitment, and unlocked by a participant that holds the secret and who can generate the relevant proof, called the nullifier. The commitment is issued off-chain by the sender of the transaction, having as secret information the value and the receiver's public key. Similarly, the nullifier is computed off-chain by the receiver, proving that he owns the necessary information to spend the locked value. The commitment and the nullifier are registered on-chain. The commitments are registered in the commitment tree, and each time a transfer is required, a nullifier for one of these commitments needs to be issued and then registered in the nullifier set as future proof for avoiding double-spending attacks. The verifiers of the nullifier are all mining nodes that need to validate the integrity of the transactions.

Homomorphic encryption is another approach that aims to improve the privacy of blockchain solutions in MimbleWimble [53], a system designed to provide an untraceable version of Bitcoin. It is a side chain that takes advantage of cryptographic properties to hide transacted values. In Bitcoin, each transaction is represented by an input amount that has an associated past UTXO that it unlocks, and an output amount locked by the receiver's key. The restriction imposed by the Bitcoin system is that the output amount of a transaction should never exceed the input amount of it. MimbleWimble uses Elliptic Curve Cryptography and Homomorphic encryption, leveraging on the fact the computations applied on the cyphertexts offer the same results as if applied on the plaintext. Therefore, in MimbleWimble the transacted amounts are blinded and by applying computations on the obtained cyphertexts they are further involved in mathematical operations proving that the input values of the transaction equal with the output values, obtaining the same results as they would have been performed directly in plaintext.

Another commonly used strategy, that hides the transferred asset amount of a transaction and the parties involved, is the ring signature. CryptoNote [54] is one of the first protocols that proposes the use of ring signatures for issuing transfers by specifying a group of possible signers to avoid the possibility of discovering the exact sender of the money. Considering a group of N parties (sub-group of network participants) involved in the ring, one of the parties can sign the message, resulting in a signature that can be verified by anyone in the network, but without the possibility of detecting the exact signing party. Furthermore, in Monero [18], the authors use Stealth addresses for the receiving parties. Each Monero account is composed of two private keys (view and spend key) and the public address. As their name suggests, the view key is used to track all the transactions that were published and are destined for that account. The spend key is used to send transactions and the public address is the one used by a sender to compute the one-time public key (stealth address), unique for each transaction. The stealth addresses offer non-linkable transactions, which means that the outputs are not associated with the addresses of the wallets. However, this solution does not provide complete privacy since the sender can trace when the money is spent by the receiver. A completely private system would need to offer both non-linkable and non-traceable transactions.

Table 2 presents comparatively the main privacy-preserving techniques identified for DLT and the privacy features they are offering. The coin mixers do not offer complete privacy, although they offer non-traceability mechanisms, the actual values transmitted are still visible. In terms of non-traceability however, the Zero-Knowledge Proofs were not yet validated as completely untraceable, due to the complexity of the algorithms involved. ZCash aims to provide traceable capabilities for the system, to be able to detect malicious users, which leads to the conclusion the transactions may not be completely untraceable [19]. Private-Public Key cryptography and Coin mixers also have one main disadvantage, because they rely on central authorities to provide Key sharing services and mixing services respectively. Another important property is the advanced scripting capability, where Coin mixers and Homomorphic Encryption mechanisms prove not to be a good solution, while the Zero-Knowledge Proof solution although feasible, has the drawback of using high computational resources for applying the cryptographic algorithms.

Table 2. DLT main privacy presenting techniques.

Features	Private-Public Key Encryption	Zero-Knowledge Proofs	Ring Signatures	Homomorphic Encryption	Coin Mixers
Hidden Data	yes	yes	yes	yes	no
Non-traceable	yes	n/a	yes	yes	yes
Non-linkable	no	yes	yes	no	yes
Decentralized	no	yes	yes	yes	no
Private Business Enforcement	no	yes	no	no	no
Transaction validation by network	no	yes	yes	yes	yes

2.1.3. Business Rules Enforcement over Transactions

The capability of a DLT system to support business implementation that can be run in a decentralized way, and then be verified and audited by all the nodes in the system, is usually provided through smart contracts. Opposed to the concept suggested by their names, the smart contracts are not very smart and may not provide a contract in the legal sense, but rather they are pieces of code similar to the stored procedures that are executed and validated by the nodes in the system, whenever they are triggered. However, there are DLT systems such as Bitcoin that are offering few possibilities for scripts to be implemented. Not being a Turing Complete language, the Bitcoin script language does not permit the implementation of complex business logic required for more advanced use cases. As a result, new DLT systems have emerged that allow customizing decentralized applications and enforcement of business logic in a decentralized way regarding when, how, and by whom may an asset transfer be executed.

In Table 3 a comparison of the main state of the art approaches for enforcing the business rules on DLT is presented. There are two types of approaches that allow smart functionality to be implemented and run across the nodes of the network: stateless and stateful [55]. The Stateless implementation is offering the possibility to implement custom logic at the level of transactions. Whenever a transaction is issued there is a set of rules that can be verified before rendering the transaction valid. The Stateful systems, on the other hand, offer Business-oriented functionalities, by focusing on the rules that govern the use case, and keeping the state of the business in tamper-resistant structures (Patricia Merkle Trees, adapted from [56]) that are easily verifiable and audited by the entire distributed system. In the Stateful Systems, the transaction has the role of triggering changes and applying updates on the stored state.

Table 3. Business rules enforcement on DLT.

Type	Implementation	Platform	Enforcement Flexibility	Costs	Exploitation Risks
Transactional Rules					
Stateless Transaction Oriented	Built-In Enforcement	Bitcoin [4], Litecoin [17] Nxt [57]	Limited Templates	None	Low Low
	Piggy Backed Enforcement	Counterparty [58]	Turing Complete	Fee per instruction	High
State Storage					
Stateful Business Oriented	Smart Contracts & Merkle Patricia Tree	Ethereum [15]	Turing Complete	Fee per instruction	High
	Smart Contracts & NoSQL DB	HyperLedger [34]	Turing Complete	None	High

In terms of functional complexity, some systems allow full computation capabilities by supporting Turing Complete languages for smart contract implementations or partial capabilities by offering a limited range of operations or fixed predefined templates. Both approaches have their advantages. On one hand, full capabilities are desired to be able to model and enforce any complex business system. Turing Completeness allows this, but it requires higher transactional costs. The on-chain computation demands for each node to execute possibly complex scripts; thus, the costs are proportional with the number of instructions. Furthermore, a complex and flexible language makes the system susceptible to different kinds of attacks that can be caused by exploiting different language shortcomings or human errors during the business logic implementations like call depth attack, race conditions, timestamp dependency, transaction ordering dependency, etc. On the other hand, the risk of exploits is highly reduced by limiting the operations allowed or by providing predefined templates.

2.2. Network Layer

The network layer technologies are related to the peer-to-peer network formed by the nodes that hold copies of the ledger (full nodes or light nodes) and participate as active players in the network. The main technological components identified at this layer are targeting the data propagation among the nodes, the peer's registration and network permission, and the consensus among peers.

2.2.1. Data Propagation and Replication

In terms of transaction data propagation, the first generation of DLT systems (Bitcoin [4], Litecoin [17], Ethereum [15], etc.) relied on full-discovery or global disclosure. This is one of the strongest features of blockchain systems since a complete replication of the data offers high availability and reliability. However, there are use cases (e.g., banking, enterprise data) that impose restrictions regarding access to transaction information [59]. Two categories of systems have been identified based on how the transactions are propagated in the system. Firstly, the global disclosure mechanism, implemented by the systems where all the full nodes have access to all the transactions published in the system, and secondly, the selective disclosure mechanism where nodes have access only to exclusive transactions that are targeting either specific businesses or only the involved parties.

Most of the blockchain ledgers adopt a global disclosure approach to offer high reliability in an open system where any node can join. The entire system is a peer-to-peer network, where all the nodes are equal. Whenever a new event is issued (a new transaction, a new block) the data is propagated through the entire network, and each node can verify and validate the integrity of the data. The redundancy in storage and computation makes it very difficult for a malicious node to influence the system to its advantage. To attack (e.g., double-spending attack) on a globally disclosed DLT, an elaborate plan must be conducted by the malicious node. It must analyze the network topology (network segmentation) and issue contradictory actions for each half of the network, with the purpose of convincing half of the network to agree with the malicious action taken.

Having a global disclosure between all the peers in the network has obvious advantages since such a system benefits from the high replication and availability brought by a large number of nodes, as well as Byzantine Fault Tolerant consensus between these nodes regarding the data. However, some clients/businesses prefer having more privacy and control over their data. This property is especially desired in private and consortium chains (e.g., banking systems), where the transactions are required to be shared only between the transacting parties. Although such a paradigm shift may lead to lower reliability in the system, the risks are highly attenuated if these requirements are implemented in permissioned systems where each stakeholder has its identity known and can be held accountable for his actions.

One of the selective disclosure approaches is presented in the Hyperledger Multichannel Architecture [60]. The system relies on third-party entities, called Orderers, which are required to order the transactions and publish them according to the category (business specific) in a corresponding channel. A Byzantine fault-tolerant consensus protocol is implemented between the Orderers, to ensure consistency between the decisions. A channel is a business-specific queue that broadcasts all the transactions to the subscribed parties. All the subscribers (peers) will receive the transactions in the same order in cryptographically linked blocks. A peer can be subscribed to more than one chain, but the chains do not interact with each other and each block received will contain only transactions corresponding to the corresponding business. Quorum [49] is another approach that aims to improve security by keeping the exclusive transactions shared only between the involved parties. The system is a hybrid between the global and selective disclosure paradigms, by allowing public transactions to be fully replicated and exclusive transactions to be shared only across the parties. The Quorum's privacy engine defines a private state tree that is updated with contracts and transactions that are sent point-to-point only to the interested parties. The private transaction contents are encrypted using Public Key cryptography, and only the users holding the private keys have access and can decrypt the actual content of the transaction. Proof of these events is also registered in the public chain, by hashing the encrypted private transaction. A similar permissioned implementation is also designed in Corda [61] where the network is formed of permission services, notary services, and peers. The system aims to provide redundancy while also keeping the transactions only known to the involving parts. Any transaction that occurs in the system must be signed and approved by both participants, and by the notary service responsible to validate transactions and prevent double-spending events. The notary service can be one entity or multiple entities that are coordinated by a consensus algorithm.

In Table 4, the comparison between the Data propagation patterns found in the literature is presented. One of the biggest disadvantages of the current selective disclosure systems is their trust in different central authorities. The Quorum system requires some level of trust between the private parties, and the other systems rely on central authorities that are responsible either for forwarding the messages like in the case of the Hyperledger MultiChannel system or on authorities responsible to validate the integrity of transactions like in the case of Corda [61] or Plasma [62]. Consequently, selective disclosure should be considered only in trusted environments, where the central authorities can be considered a source of truth, while for public environments, global disclosure should be considered such that any party involved in the network can validate the integrity of the transactions.

Table 4. DLT data propagation patterns.

Type	Platform	Trusted Parties	Global Disclosed		Selective Disclosed	
			Data	Structure	Data	Structure
Public DLTs	Ethereum [15], Bitcoin [4], etc.	-	All transactions	Blockchain	-	-
Business Specific Chains	HyperLedger Multi-channel [60]	Orderer	-	-	Exclusive transactions	Queues, Blockchain
	Plasma [62]	Central Authority, N delegates	Public Transactions + settlements	Blockchain	Exclusive transactions	Blockchain
Point-to-Point transactions	Corda [61]	Notary Service	-	-	Exclusive transactions	Local database
	Quorum [49]	Private parties	Public transactions, Hashes of Exclusive Transactions	Blockchain	Exclusive transactions	Merkle Patricia Tree

2.2.2. Permission Mechanisms

Over the years the private institutions that realized the potential of the systems behind DLT, started to evaluate the integration of such systems with their businesses. However, some key components rendered the public chain unsuitable for many institutional and enterprise solution requirements, so they started to investigate new systems that address the issues regarding the governance and the permissions of the system. Firstly, in an enterprise solution, the participants need to be known and vetted before given access. Such a decision has a great impact on the system, even in terms of security and consensus. Since the participants are known, thus can be held accountable for their actions, the need for a high energy-consuming algorithm like Proof-of-Work is no longer justified. Therefore, there is a strong relationship between the requirements regarding access rights and the consensus algorithms suitable for a specific business.

The difference between public, private, permissionless, and permissioned DLT/blockchain is given mainly by the rights of the users in the system. Based on the classification presented in Table 5, the difference between private and public chains is established according to the target audience that has access (reading rights) to the chain. Restricting the access of a group to the chain renders the chain private. According to the group of people accessing the chain, it can be a consortium or an enterprise solution, where the consortium solution operates under the leadership of a group of companies, and the enterprise solution is under the operation of a single entity.

Table 5. Public vs Private Blockchain permissions.

Action	Public Chain		Private Chain	
	Permission-Less	Permissioned	Consortium	Enterprise
Chain Access	Everyone	Everyone	Group Owner	Group Owner
Transactions	Everyone	Owners & Validated Users	Owners & Validated Users	Administrator
Commit to chain	Everyone	Owners & subset of Validated Users	Owners & subset of Validated Users	Administrator

In public DLTs, some restrictions can also be imposed regarding the users' access and permissions. In a permissioned ecosystem, the validators are known and accountable for their actions, thus a certain level of trust between the nodes can be considered. In a permission-less system, on the other hand, any user can perform any type of action (transactions of an asset, as well as commits of new blocks to the chain). Consequently, permission-less DLTs require Byzantine Fault Tolerant consensus algorithms, since the openness of the system allows even malicious nodes to join, making the network susceptible to a larger range of attacks.

2.2.3. Consensus Protocols

Figure 3 presents a taxonomy of the consensus algorithms which are classified in Non-Byzantine fault-tolerant algorithms and Byzantine fault-tolerant algorithms. The difference is given by the ability of algorithms to reach an agreement, integrity, and termination in case of existing faulty or attacker nodes in the distributed system, thus Non-Byzantine fault-tolerant ones rely on the assumption that all the nodes are fair, while the Byzantine fault-tolerant algorithms can handle situations when the number of malicious nodes is as high as half of the total number of nodes.

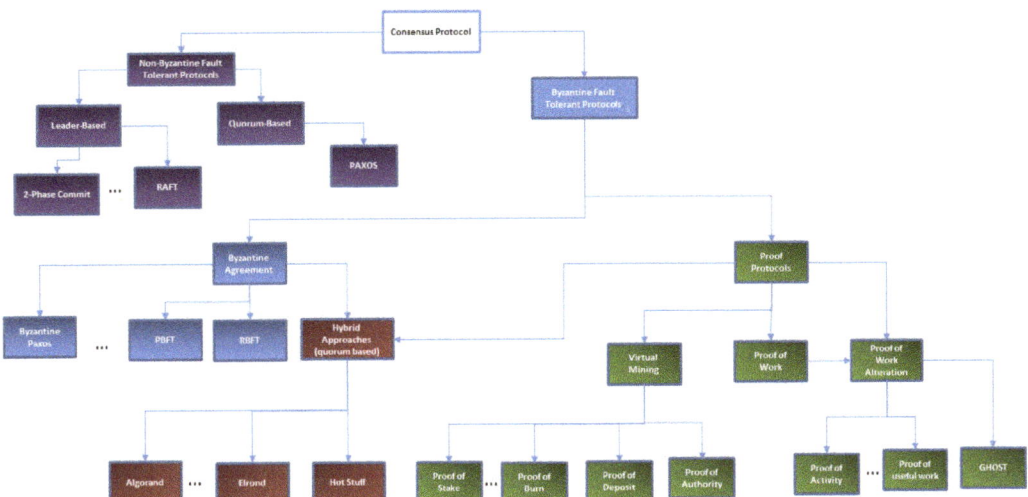

Figure 3. Consensus algorithms taxonomy.

The *Non-Byzantine fault-tolerant protocols* are leader-based, such as 2-Phase Commit [63] and RAFT [64], where a leader election algorithm is used to select a leader that will centralize the votes and commit the transaction. Furthermore, they are quorum-based, where a subset of the processes is selected to validate the transaction using a voting scheme. A well-known algorithm of this class is the Paxos algorithm [65] that solves consensus in a network of processes that may fail but are correct (there exist no faulty processes that may lie). In Quorum, RAFT algorithm is used, where a predetermined leader is creating a block that is sent to each node in the cluster [66].

The *Byzantine fault-tolerant protocols* aim to assure that the peers can agree on a system valid state even in case some of them feature faulty or malicious behaviors. The idea is to find a model and protocol for a network of message-passing processes, some of them being faulty, such that a general agreed state can be extracted from the distributed system. The Byzantine fault-tolerant protocols can be classified as Byzantine Agreement protocols and Proof Protocols [67]. In terms of finality, proof-protocols are known not to be final, however, they offer probabilistic finality, since once many blocks are sealed over, the probability of a block's state to change is very low.

The *Byzantine Agreement (BA)* protocols use a quorum-based mechanism where a subset of the nodes must agree on a transaction validity. Examples of such algorithms are the Byzantine Paxos algorithm [68], the Practical byzantine fault tolerance algorithm [69], and variants that address the robustness such as Ardvark [70] and RBFT [71] or that address the performance problems of PBFT, such as Q/U [72], HQ [73], Zyzzyva [74] and ABsTRACTs [75]. An interesting Byzantine fault tolerant distributed commit protocol is proposed in [76], where the authors enhance the classical 2-Phase Commit protocol by replicating the coordinator to successfully terminate when the coordinator failed and by building a quorum of coordinators to validate transactions and identify malicious participants.

The *Proof Protocols (PP)* are used by most of the public DLT systems [77,78] for supporting the consensus mechanisms to ensure the consistency of the ledger state across the network nodes. The Proof Protocols have defined two categories of nodes: Provers and Verifiers, where the Prover who may have unlimited resources needed to convince Verifier nodes with limited resources, about the truthfulness of a statement. As opposed to BA protocols, which use a quorum of participants to validate a transaction by voting, PP such as Proof of Work (PoW) and various alterations algorithms validate a transaction (or a set of transactions) by solving a computationally-intensive problem by a Prover that requires a lot of physical resources and makes infeasible for an attacker to cast an erroneous vote. The time needed by the prover to solve the computationally intensive problem gives the mining rate and directly influences the throughput (number of transactions) and the network scalability. From the initial implementation of PoW in Bitcoin, where one block was generated every 10 min, PoW variations have been proposed aimed at improving the mining rate to obtain a higher throughput of transactions per second. The Greedy Heaviest Observed Subtree (GHOST) protocol [79] proposed by Ethereum increases the mining rate from 1 block per 10 min to 1 block per ~15 s. To avoid the potential problems that may arise due to delayed propagation of blocks, GHOST uses references to orphan blocks or uncles (valid blocks that were not accepted in the main chain due to network delays) to increase the weight of the longest chain. In this sense, each new block can contain references to previous uncles and for each of the referenced uncles, the miner will receive a small incentive, consequently, the miner of the uncle will also be rewarded when a new block refers to it. This mechanism discourages the faulty miners to mine on forked chains and from perusing long-range attacks. Other variations of PoW have been considered to impose some restrictions on the hardware devices used for mining by encouraging the implementation of ASIC (Application Specific Integrated Circuits) resistant algorithms for hashing. This came because of Bitcoin's early years when hardware companies started to profit from the popularity of blockchain solutions by developing ASICs to increase the hash rate of the computing nodes. However, one such circuit may cost around 3000 dollars [80], which makes it unprofitable for a simple user to invest in such hardware and gives more power and control to large companies and the manufacturer. To avoid this problem, the next generation of DLT solutions researched and applied new hash functions that are ASIC resistant. ASIC resistant algorithms try to shift their strategy from CPU intensive algorithms to memory intensive algorithms, called Memory hard puzzles. This came because the performance of processors has increased over time at an exponential rate, as opposed to the memory which has known a more linear increase. The purpose of these algorithms is to design a method that requires large amounts of data to be stored, that cannot be efficiently parallelized. Scrypt [81] is one of the first ASIC resistant algorithms and is currently widely used by many applications. However, Litecoin, which is one of the top platforms that use this algorithm set the memory size at 128 KB [82] thus making it possible to be stored at the CPU cache level. This restriction was applied since the Scrypt algorithm requires the same resources for solution verification as for the solution discovery and higher requirements would stress too much the regular non-mining nodes. Dagger Hashimoto [83] on the other hand, is an algorithm that provides an easy verification solution, thus allowing the Prover's requirements in memory size to increase up to 1 GB RAM. Equihash is also a widely used hashing algorithm. However, the main disadvantage, as the authors themselves state [84], is that the algorithm is parallelizable, which is not a quality desired in ASIC resistant algorithm. Finally, the Cuckoo hash cycles [85], used in [86,87], are also considered a reasonable solution when talking about ASIC resistance. Other relevant variations of PoW algorithms aim at giving a purpose for all the energy and computational resources of the network [88]. Since the network uses large computational resources whose only purpose is to prove and validate the next block of the blockchain, the concept of Proof of Useful Work is launched as an alternative to trying to use the computational power for a publicly beneficial domain. Such implementations aim to do research work (or Proof-of-Research). They gather the computational power across the network to provide solutions to some of the world's problems. CureCoin [89] is implementing an algorithm called SigmaX that aims to perform protein unfolding to find a cure for different diseases. Proof of Activity [90] is a PoW alteration algorithm found in Decred [91]. The algorithm starts as a simple PoW algorithm until one correct hash is found; the block is then transmitted in the network, but it is not yet added to the blockchain. To become a valid block, it needs to be signed by N holders in the network. The PoW obtained hash is used to generate N numbers that correspond to N coins generated since the genesis of the blockchain. Each of these coins has one current stakeholder who will be required to sign the current block. The signature of all the N stakeholders is required to consider the block valid. In case that some of the stakeholders are not online and cannot sign, then the miners will continue their job to find a new hash and ask other stakeholders to sign the block. This approach makes attacks upon the network more difficult since it makes use of the advantages brought by both systems.

Virtual Mining Protocols offer an alternative to the PoW by keeping a high cost for the Prover, but changing the resource consumed. If the cost of the Prover in PoW is the energy consumed, which would be lost if the Prover does not offer honest work to be validated and rewarded by the network, in the virtual mining Protocols the cost is a deposit of coins that are offered as insurance for their honest work. If up until now the node was chosen based on its result to the computationally intensive problem, now the node will be elected in a pseudo-random way, and the chance of winning will be proportional to the number of coins/stakes of the owner of the system. Thus, in Virtual Mining Protocols, the clients have the mining potential proportional to the percentage of the stake they hold. Four virtual mining approaches have been identified across different solutions: Proof of Stake considers the age of the coin in the algorithm, thus requiring for some coins not to be spent for some time; Proof of Burn requires a relevant amount of coins to be destroyed and a proof of the destroying transaction to be provided; Proof of Deposit requires for some coins to be put away for some time in a vault; Proof of Authority suggests that only trusted parties are entitled to provide commits to the system, which can be required where high-security properties need to be implemented [92], like in the case of private Enterprise solutions. However, all four algorithms have the same purpose that is, incentivizing the honest work of the miner by promising as a reward a sum of coins greater than the initial insurance. According to [93], the Casper version of Proof-of-Stake (PoS) is considered a suitable alternative for the permissioned systems, by considering only a fixed set of users as validators of blocks. Another flavor of Proof-of-Stake commonly used for permissioned systems is the Delegated Proof of Stake (DPoS). In DPoS, N witnesses are periodically selected by stakeholders of the system, such that enough decentralization is ensured. Out of the N witnesses, each witness has its chance to propose the next block, and then be rewarded for its contribution. From existing Virtual Mining Protocols, the PoS a good potential of becoming the most used consensus protocol in DLTs because it addresses fundamental problems of the PoW protocol such as computational waste and high-power demand [94]. Anyway, in the case of the PoS algorithm, since the nodes propose a new block by guaranteeing with their stake it gives rise to the "nothing-at-stake" vulnerability. This means that when a fork appears in the context of a network partitioning, an attacker node can propose a block on either chain, hoping that at least one block will be accepted. The node guarantees each proposed block with its stake, but due to network partitioning, it is difficult for other nodes to observe and penalize this misbehavior. This situation can lead to other forks or to the fact that the attacker node receives rewards for proposing new blocks. In PoW algorithms, the "nothing-at-stake" vulnerability is avoided since when proposing a new block, the node has to solve a computational puzzle that consumes electrical energy, and by proposing two blocks on two chains from a fork means that the node has to solve twice the problem, thus doubling its costs. There are two categories of PoS mechanism: (i) chain-based PoS that mimics PoW by assigning pseudo-randomly the right to generate new blocks to various nodes and (ii) Byzantine Fault Tolerant PoS that is based on BFT research. They address the "nothing-at-stake" vulnerability in different ways. The chain-based PoS are penalizing nodes when sending multiple blocks on competing chains (e.g., Slasher [95,96], or Casper [93]). The BFT PoS mechanisms allow validators to vote on blocks by casting several messages, with two rules: finality condition (to determine when a hash is finalized) and slashing conditions (to determine when a validator misbehaved and must be excluded). A block is considered finalized once enough votes have been cast and all nodes from the DLT agree on adding it to the canonical history. This involves sending many messages in the network to make aware other nodes that a new block was proposed and running a version of the Byzantine Agreement on the new block.

Propagating many messages in the network impacts system scalability, thus methods to reduce the number of messages exchanged are needed leading to the development of hybrid approaches between Byzantine Agreement and Proof protocols [97]. Two techniques are found in the literature addressing this: (i) quorum based voting—when a node is selected randomly as the prover and a subset of nodes are selected to be verifiers that run a Byzantine Agreement protocol (Algorand [94]); and (ii) sharding-based approaches—where the blockchain is split into shards for inter-shard transactions and only transactions that involve nodes from two different shards need message propagation between shards (Elrond [98]). Algorand is based on a new and fast Byzantine Agreement Protocol used to generate a new block through a binary Byzantine Agreement (BA*) protocol that enhances the traditional BA protocol to work in rounds in a synchronous environment with at least 2/3 players being honest. Furthermore, cryptographic sortition based on Random Verifiable Functions is used to select a subset of the users to be members of the BA* algorithm. A cryptographic function is used to select a new leader based on a previous block. The leader will be in charge to propose the new block. A set of verifiers is used to check the validity of the new proposed block. The choice of the leader is not predictable, thus making it impossible for an attacker to alter the new block. Furthermore, leaders learn of their role without informing others only after proposing the new block, thus avoiding attacks. After a new block is proposed, the leader has no importance for the algorithm. However, the verifiers must agree on the new block, and they run the BA* algorithm in rounds, at each step players being replaced, thus avoiding cases when many verifiers are corrupt. Elrond is based on a sharding approach, splitting the blockchain and account state in several shards where parallel validation can occur using a consensus algorithm based on a secure PoS. The consensus algorithm follows a similar approach as Algorand with a prover and a set of validators chosen randomly within a shard and running a Byzantine Agreement algorithm to validate the proposed block. Finally, Hot Stuff [99] proposes a consensus algorithm using a leader-based Byzantine fault-tolerance protocol for partially synchronous distributed system models where a chosen leader drives the consensus decision at the rate of the maximum delay allowed by the network.

3. Scalability Tier

The DLT scalability limitations are mainly driven by the restrictions imposed by the Protocol and Network Tier solutions (e.g., the consensus algorithms). To achieve higher scalability, one option would be to curtail some of the features of Protocol and Network Tier. This can be done either by compromising the security, the immutability, or the consensus of the DLT. Because most of the time this is not acceptable, the scalability challenges are open for research for all DLT variations. In this sense existing concepts such as distributed databases or file systems, have been reconsidered and integrated with Protocol and Network Tier, to allow the implementation of solutions for the Scalability Tier [7,100].

Anyway, due to the advent of blockchain platforms and applications, most of the nowadays literature is focused on the scalability limitations of this type of ledger. They are imposed either by maximum block size (e.g., 1 MB Bitcoin) or by a cost constraint (e.g., gas consumption and gas price in Ethereum). These constraints are combined with the strict periodicity of the block generation (e.g., 10 min for Bitcoin, 15 s for Ethereum) imposing limitations in the number of transactions processed. Moreover, they are impacting both the storage and the processing capabilities (e.g., due to the gas consumption costs in the case of Ethereum smart contract execution). Bitcoin reportedly can allow 7 transactions/second on average [101], while Ethereum registers 13 standard transactions/second or 7 transactions/second in case smart contract execution is involved [102]. Even private deployments reach certain limitations. Hyperledger is advertising 100,000 transactions/second, although reports show a lower limitation of 700 transactions/second [60].

3.1. Storage Size

With the increased storage capabilities of the systems, much of the paper documentation has been digitalized in domains like healthcare, intellectual properties, real estate, legislative contracts, etc. Furthermore, the media and social network use cases are more and more flexible, providing increased storage options for users to store their files (documents, photos, videos, etc.). DLTs caught the interest of these domains, aiming to maximize their potential, by ensuring immutability (legislative and real estate), provenance (intellectual properties), security (healthcare), etc. However, the greatest challenge of integrating DLT solutions with these domains is limited storage capabilities.

To improve the storage scalability several solutions have been proposed, such as Sharding or the integration of well-known file systems with existing DLTs. They are aiming to store all the data outside the DLT and keep only a digital fingerprint of the data on the Protocol and Network Tier. While the data kept on the Protocol and Network Tier benefits of all the advantages the system provides (consensus, immutability, security, etc.) it is considered a source of truth in the validation of data that is stored on the Scalability Tier.

Sharding is a solution implemented to improve storage scalability [103]. Different nodes are assigned to process and store only a corresponding sub-category of transactions [104]. A simple sharding technique is to split the network in shards corresponding to the transaction's prefix: 0×01 shard, 0×02 shard, etc.

For example, in the sharding mechanism proposed by Ethereum Sharding [105], the system defines objects at three different levels: level 0—transactions; level 1—collations; level 2—blocks. The collations are the data structures responsible for package transactions that belong to a shard. The collations are created and sealed by Collators that are nodes in the network registered on the main chain in the Validator Manager Contract. The Collator deposits a sum of coins on the main chain based on which they will be chosen in a Proof of Stake manner to validate the next collation. The header of the proposed collation will then be verified on-chain and added in the next block on the main chain. Cross-Sharding communication is also possible by providing Merkle-Proofs of existing transactions from the main chain. Similar approaches are investigated by Elrond Network [106], Hyperledger [34,60], Elastico [107], Omniledger [108] and Rapidchain [109].

The Scalability Tier solutions that use file systems as storage mechanisms allow large files to be stored by fragmenting, encrypting, and sharing chunks of the original file between the nodes, while the hash of the original file is stored in the Protocol and Network Tier. The nodes storing the data need to respond to periodic checks regarding the integrity of the stored data, and a reward scheme is implemented for their services.

Figure 4 shows an example of storage mechanism and integration with the Protocol and Network Tier in the case of blockchain ledgers. There are several successful implementations of such distributed file systems among which, worth mentioning are: Storj [110], IPFS [111], Filecoin [112], MediaChain [113], Decent [114], Sia [115], MadeSAFe [116], Swarm [117] and Arweave [118].

IPFS (InterPlanetary File System) is one of the most used Scalability Tier solutions for file storage. In this case, when a user publishes a large file using its own IPFS node, the node will first fragment the file in smaller chunks, the hash of each chunk becoming a node in a Merkle DAG, whose root is the hash of the initial file, thus making use of hash pointers to ensure tamper-evidence. For security reasons, the chunks stored are of standardized sizes, so that an attacker cannot extract any useful information by analyzing the size of a chunk. The owner of the data is responsible to hold the private key used to encrypt the chunks of data that are scattered across the network. This makes the system highly secured since even the data is stored across multiple nodes, the mechanisms make it impossible for anyone holding the data to use it since it is encrypted and fragmented. Moreover, it ensures security through encryption and no downtime since the file is shared across multiple users. The system offers the possibility to transfer data, check the availability and the integrity of the stored data, retrieve the data, and pay for the service provided.

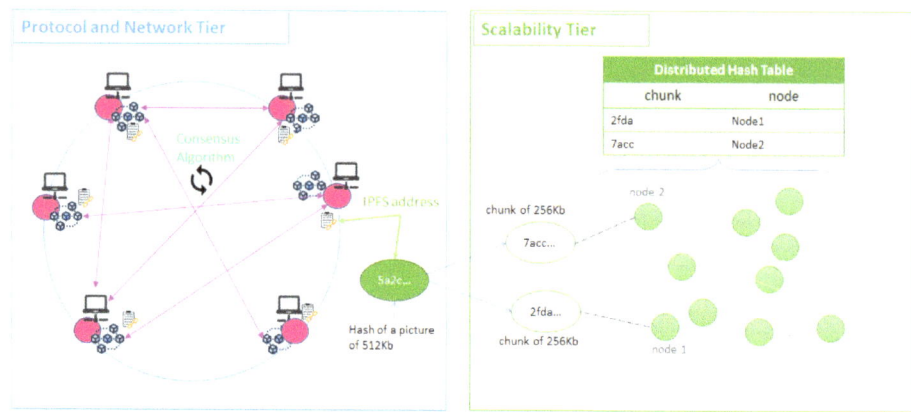

Figure 4. Scalability Tier—File System Storage Mechanism.

Similar implementations such as Storj [110] and Filecoin [112], are proposing to reward and motivate the decentralized nodes to act honestly regarding their storage services. Ethereum Swarm [119], is a peer-to-peer system that aims to store data in a decentralized way and relies on immutable content-addressable data. While IPFS needs Filecoin to validate storage proofs, Ethereum Swarm proofs are validated at the contract level and rely on incentive schemes based on the native coin, Ether.

In Table 6 a comparison between the identified storage scalability solutions is presented. Sharding presents a promising alternative to the classic DLTs, by providing increased storage capabilities. Using sharding the DLT storage capacity is multiplied with the number of shards, having the block sealing process parallelized with each shard. However, by increasing the number of shards, fewer nodes get to be assigned per each shard for validation. This can easily make the network susceptible to attacks since by attacking one shard the entire system can be compromised. The file systems solution even if it provides great scalability in terms of storage, also requires a degree of trust between the storing nodes. For example, in the case of IPFS, since it is not fault-tolerant on its own, the DLT storing the hash ensures only tamper-evidence in the system but does not make the system tamper-resistant. Each time an update is applied to one of the files, the hash pointer changes as well, requiring a transaction updating the entry on-chain as well. While this is desired to keep a tamper-evident system, a high frequency of updates will also lead to higher costs.

Table 6. Scalability tier storage solutions.

Features	Protocol & Network Tier	Scalability Tier	
	Fully Replicated	Sharding	File Systems
Immutability	Yes	Yes	No
Trusted Parties	None	None	Peer nodes
Byzantine Tolerant	Yes	A tradeoff with the no. of shards	No
Storage Scalability	Low	Medium	High
Cost	High	Medium	medium

3.2. Transaction Throughput

The number of transactions processed by blockchain ledger is important for implementing decentralized applications where micro-payments should be exploited (e.g., Energy Sector, Media Services, etc.). Micro-payments are online transactions involving small amounts of money. These small amounts of money are often used in exchange for different goods or services, and most of the time require many transactions over a period. The problems that arise by integrating the micro-payments are the high cost accumulated as a result of the mining fees paid for each transaction and the congestions problems at the level of the Protocol and Network Tier due to the small size of the block. The most promising solutions analysed are the Sharding, Sidechains, and Payment Channels (see Table 7).

Table 7. Transaction scalability solutions.

	Protocol and Network Tier		Scalability Tier	
	Fully Replicated	Sharding	Sidechains	Payment Channels
Trusted Parties	None	None	Depending on implementation	None
Transaction Scalability	Low	Medium	Medium	High
Cost	High	Medium	Medium	Low

Sharding can be considered a suitable solution for increasing the transaction throughput as well as the storage [120]. Both improvements come because of introducing clusters of nodes responsible for specific categories of transactions. Higher transaction throughput may be provided by delegating the transactions to a different category of nodes and parallelizing the validation of these transactions. An issue arises when increasing the number of shards and transactions. When a transaction is sealed by a shard, it may reference a transaction that is not in the log of transactions of that shard. As a result, each time such a transaction needs to be validated, a cross-shard communication is required to issue Merkle Proofs of the referenced transaction. Consequently, increasing the number of shards and transactions will also introduce a communication overhead that may impact the scalability of the solution. More exact evaluations of the overhead introduced will be possible only after these currently researched solutions will offer a full specification and deployment on the public networks.

Sidechains [121,122] are proposed as alternatives for increasing the scalability of blockchain ledgers. A side chain is processing transactions in parallel with the main chain. The transactions of the sidechains are always rooted in a locking transaction in the main chain. Once proof of a locking transaction is made on the side chain, the actors may start using the assets by transacting on the side chain. To return to the main chain, proof of the latest state from the sidechain must be made to unlock the coins on the main chain. There are different reasons for connecting to side chains: testing new functionalities, extending the main chain functionalities (e.g., RootStock [123] enabling smart contract execution), or moving business-specific implementation to another less expensive chain (e.g., Plasma as a tree of sidechains [62]). In either case, moving part of the transactions from the main chain to side chains can lead to an improvement in the transaction throughput.

The sidechain implementation offers an alternative to sharding by fully relying on the information stored on the side chain and integrating with the main chain only when locking and unlocking the coins. This solution may offer some improvements to the underlying chain, by taking over some of the transitioning load. However, from a scalability perspective, the transaction throughput is still limited since the side chain is most of the time a blockchain ledger with the same constraints as any Protocol and Network Tier solutions. Additionally, some issues regarding the overall security of the systems need to be considered. Upon returning to the main chain, the transactions validated by the side chains are valid even if the validators' network of the side chain differs from the ones of the main chain.

The Payment Channels mainly implemented in Lightning Network solutions [124] are one of the best choices for improving transaction throughput. The mechanisms of the Lightning Networks were firstly defined for Bitcoin in [124], and future implementations have followed for other networks: Raiden for Ethereum [125] or Bolt for Zcash [126]. The Payment Channels aim to combine all the small payments into one large payment at the end of a service period. The Lightning Network provides a point-to-point network that runs on top of the blockchain network as depicted in Figure 5.

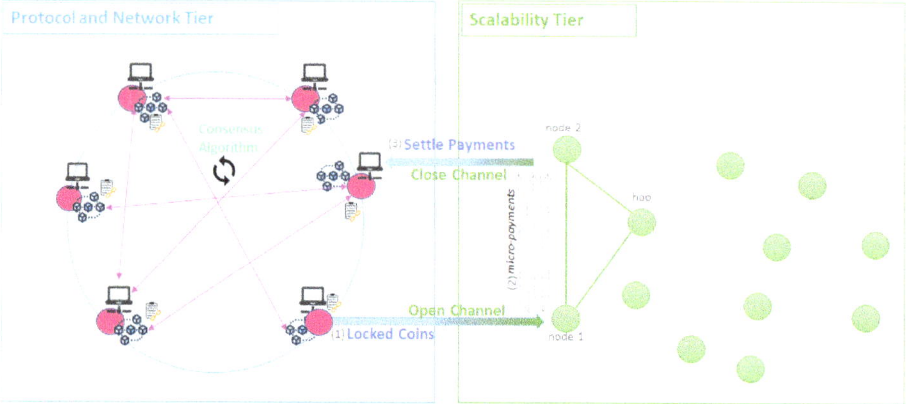

Figure 5. Scalability Tier—Payment Channels Mechanism.

It relies on hash time locks and cryptographic secrets to ensure the reliability of off-chain payments. The nodes in the Lightning Network exchange cryptographic signed transactions among them. They represent the payments exchanged between nodes offline. At any point, the payments can be deployed on the main chain to securely redeem the associated coins. Whenever two parties aim at exchanging many micro-transactions, the first step is to open a channel between them, represented on the blockchain as an opening transaction that locks the maximum amount of money that the two parties could transact off-chain. The transaction opening is signed by both parties involved. To unlock it both parties need to agree on the spending of the amount. Afterward, the parties will continue to exchange messages off-chain, namely, commitment transactions, which are designed as fail-safe mechanisms against cheating off-chain. Whenever a transfer occurs, determining a change in balances, each party creates one commitment transaction specifying the updated balances, signs it, and sends it offline to the other party. Upon receiving the commitment any party can successively sign it and publish it on-chain to redeem the coins or can wait and make other off-chain transfers and return only with a future commitment transaction on-chain. To prevent any party to return to the chain with a deprecated and more favorable commitment transaction, secret-locking and time-locking mechanisms are incorporated. They provide a fail-safe mechanism such that any party acting fraudulently risks losing all the money in favor of the other channel party.

The benefit of transferring over a route is the reduction of the cost associated with the opening transactions required on-chain each time a channel is required with a specific party. For sending the transactions through a path that requires intermediary parties (hop nodes) to route the messages, additional security mechanisms are required to ensure the correct delivery of the transacted assets. For successfully issuing a routed transfer, a commitment transaction is exchanged between every two parties involved in the path channels. Hash Time-Locked Contracts [127] have been defined over the previously presented mechanism, to prevent the intermediary to unlock the routed commitments before the receiver can confirm the payment. Upon confirmation, a proof is issued by the recipient and sent to the intermediary to unlock the hash-locked commitment. If the proof is not provided during an established period (time lock) then the intermediary will no longer be able to claim the payment.

The Payment Channels offers the best scalability solution. It is implemented without relying on any third parties and provides higher transaction throughput, promising millions of transactions per second. However, once the network has many users it is more difficult to find a viable route between two parties. The most desired topology of the lightning network would be a completely decentralized network. In this case, each node has connections with other nodes, together forming a mesh of channels that are part of a connected graph. However, to be able to route money from one point to another, all the nodes from the path need to have at least the same amount of money as the one requested by the initiator. Taking advantage of this issue motivates some actors of the system to open channels with a larger number of parties and fuel the channels with enough money to be able to route and connect different parts of the network, acting like large routing hubs in exchange for small fees.

3.3. Processing Capability

The processing capability limitation is extremely relevant for blockchain distributed ledgers that allow the implementation of complex functionality and computations through smart contracts. For example, in Ethereum, the concept of gas was introduced to measure the amount of computational effort. When running smart contracts, each executed operation and processed byte of data is paid for with gas. This mechanism prevents an attacker from running extremely long tasks or infinite loops since the attacker would need to provide enough reward to incentivize the miner to execute each operation. When the reward provided runs out, the computation stops, and the transaction is dropped, thus avoiding situations where the nodes become unavailable due to attacks or complex computations. By integrating the Protocol and Network Tier system with external services, this shortcoming can be solved.

The nowadays solutions for addressing complex computational problems in blockchain ledgers are Oracles [128] and Proof of Computation mechanisms [129].

The Oracles are mechanisms that provide a secure connection between the blockchain and the outside world. They act as a trusted third-party entity, or a network of entities, for the Protocol and Network Tier. The Oracles can be used to offer results from different URLs, [130], IPFS, or units responsible for running more complex algorithms. The problem with interacting with the Oracles, directly from the chain, is that the response must be the same across any number of requests issued by the nodes during mining. This proves to be almost impossible when accessing dynamic changing data regarding weather, stock prices, etc. One problem that can appear in the Oracle-based system is data tampering or man-in-the-middle attack. In this sense, the Oracles are responsible to ensure the authenticity of data through authenticity proofs. One problem that persists is the centralized nature of the Oracles. The mechanism is presented in Figure 6 for a blockchain ledger [131]. An event containing details about the request is issued from the blockchain and intercepted by the Oracle. The necessary information is retrieved from external services and published back on the chain through a callback transaction.

Other implementations aim to outsource computing-intensive problems to off-chain nodes by implementing a Proof of Computation mechanism. Compared to the Oracles the Proof of Computation is a better choice. It is implemented without relying on any trusted party and provides validation of the result implemented directly on-chain. The proposed solutions show great potential for use cases requiring security and correctness validation for more complex computations than the ones that can be handled on-chain.

Figure 7 shows the Proof of Computation mechanism of TrueBit [129]. It relies on Ethereum smart contracts and gives the possibility of peers to request solutions for complex computational tasks. A Solver that has enough computational resources will run the tasks outside the chain and submit its results.

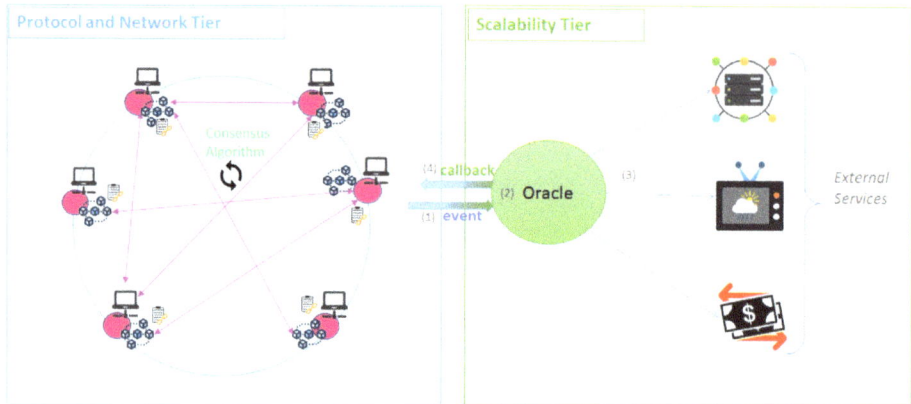

Figure 6. Scalability Tier—Oracle for External Services Integration.

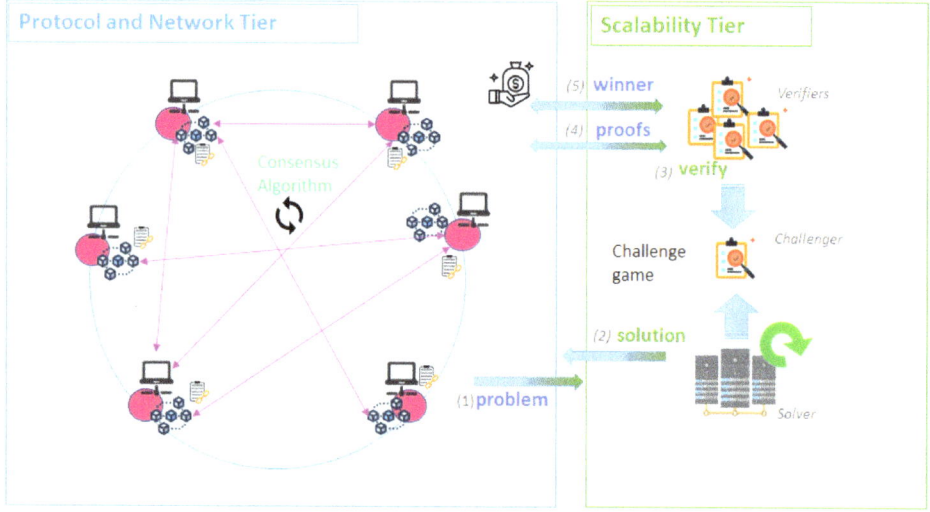

Figure 7. Scalability Tier—Proof of Computation mechanism.

Several Verifiers peers can evaluate the results, and if a disagreement occurs, a Challenger can contradict the result published, by starting a Challenge Game. Several rounds of proofs are registered on the chain to check whether the computation was done correctly. In the end, the winning part will receive a reward for its cooperation, while the part proven to be wrong will be charged for its actions. Another set of actors are de Judges that given the proof of the solution can easily verify the correctness of the game. However, a big drawback of the system is that it is limited to running tasks written in WASM [132]. Enigma [133] proposes a similar concept, of outsourcing the computation but with an added layer of privacy over the data and computation performed. By leveraging on secure multi-party computation, the proposed solution distributes the data across several nodes for computation. As a result, no central node will have access to the entire problem or solution, but to a seemingly unintelligible part of it. The proposed solution is not yet released in production, but it is currently tested on the Ethereum test network [134].

4. Interoperability Tier

By launching various domain-specific DLT applications for public use, each addressing different requirements, one problem that arises is the need for an interoperability protocol or mechanism. It should facilitate the transition from isolated DLT applications to networks of integrating DLT applications. Such a network could further unlock the potential of the DLT by allowing different types of businesses or applications to leverage on a different type of DLTs. Moreover, they will coexist and cooperate by sharing their ledgers.

The state-of-the-art solutions for DLTs integration are mostly referring to the blockchain ledgers [7]. Due to the development of a high number of blockchain platforms and applications, their interoperability is the main technological trend in the next years [135]. The most popular ones such as Ethereum, Bitcoin, or Hyperledger use different data formats and data interchange solutions making their integration difficult [136]. For example, Interledger [137] developed a protocol for allowing communication across different blockchain ledgers by using payment channels. The mechanism is called "AtomicSwap" and uses routes existing in the Lightning Network. Considering a hop that has Receiver and Sender channels opened on two networks, such as Bitcoin and Ethereum. Upon a transfer, the hop can agree to update its Bitcoin balance from the Receiver channel, in return for Ether on the Sender channel.

Most state-of-the-art interoperability solutions are addressing the transfer of coins from a parent blockchain ledger to a secondary one. Two-way peg systems [122] are mostly used for this type of Ledger-to-Ledger communication. The transfer is achieved by temporarily locking some coins on the parent blockchain and then unlocking the same amount of coins on the secondary blockchain. The transacting process is executed on the secondary chain until the user decides to return to the main chain. This is possible by repeating the same process on the secondary chain, thus locking the coins on the secondary chain, and then releasing them on the main chain. A central exchange that needs to have access to both chains is proposed as a solution for implementing the transfer of coins between two blockchain ledgers [123]. A user sends a request from the main chain specifying the number of coins and the address that should receive the coins in the secondary chain. The exchange can then simply send the same amount of coins to the specified address on the secondary chain. The central exchange needs to be a trusted entity; otherwise, the money could be easily stolen from the two chains. To address this issue a MultiSignature scheme can be used [138]. In this case, any transaction must be approved by N out of M participants, instead of relying on only one entity. It still relies on a middleman, but the risks are significantly reduced.

The entangling of the chains is presented as a potential solution for coin transfer between blockchain ledgers. It uses a secondary chain to monitor the main chain and includes all the block headers in the main chain blocks [139]. As a result, blocks from the secondary chain will have two parents: the previous block from the secondary chain and the last mined block on the main chain. The main disadvantage of this approach is that the main chain needs to create blocks at a lower rate than the secondary chain. An entangled model is used by BTC-relay that connects the Bitcoin chain with the Ethereum chain [140]. After every 10 min, proof of the last mined block in Bitcoin is provided in the Ethereum smart contract. Each time a user wants to prove the validity of a Bitcoin transaction on the Ethereum chain, it only needs to provide the Merkle Path of the transaction and the block it is contained in (a Simplified Payment Verification (SPV) proof [141]).

The sidechain solution for coni transfers aims to eliminate the middleman. The interoperability protocol of the chains implements a new way of unlocking coins, which is *proof of a locked transaction on the other chain*. The transacting parties will submit on the sidechain, an SPV proof of the transaction deployed on the main chain. The SPV proof will contain the Merkle path for the transaction submitted and mined and the hashes of all the blocks that followed. Upon receiving the proofs, the sidechain will enter a reorganization phase. This phase aims to provide the necessary time to avoid any possibility of a double-spending problem. In the case of a fork in the main chain network, anyone can offer a new SPV proof that contains the same block as the one provided initially, but without the transaction in it. If the proof provides a longer list of block hashes that confirm the missing transaction block, it is concluded that a double-spending attack was committed on the main chain and the original transaction is ignored in the sidechain as well. Upon withdrawal on the main chain, the same process occurs. The main chain lockbox is defined as a new type of transaction that can only be unlocked using SPVs of locked transactions from the other chain. The system still has a notable drawback since the main chain needs to rely on the integrity of the sidechain. If the miners of the sidechain are not honest and coins are stolen, the main chain has no way to prove the honesty of the requests if a valid SPV is provided. Furthermore, if the amount of coins is split on the sidechain, to retrieve the coins on the mainchain all the owners from the sidechain must provide the SPV, thus no partial consumption of the coins on the main chain is permitted. Drivechain [142] is an improvement of the sidechain. The protocol is like the sidechain one until the step requiring withdrawal from the sidechain back to the main chain. At this point, the SPV proofs are not used directly to withdraw the coins on the chain. During the reorganization period, several withdrawal proofs are joined aiming to consume a given amount of coins from the main chain. The withdraw transaction id, (not the actual transaction) will then be mined in the next block on the main chain's transaction. The mined ID will be interpreted as the intent of spending the locked money but will give time (1008 blocks) for all the users from the sidechain to validate the transaction and give chance to the miners to vote whether the actual transaction should be mined on the mainchain. In this way, the miners of the sidechain are prevented to commit illegal transactions, since any commit to the main chain will first be evaluated by the corresponding actors and most miners. A hybrid model of Drivechain is proposed by Rootstock [123], implementing a combination of sidechain and multi-signature federation. It uses sidechain functionality for passing coins from the main chain to the secondary chain. However, when returning not only the miners have the right to vote but also specially delegated notaries that vouch for the integrity of the transactions.

The communication between blockchain ledgers is of much interest for Ethereum as well. A solution in this platform case is to offer different chains per application. CryptoKitties [23], is such an example of an application build on an alternative chain. It is completely independent of the state and data stored on the Ethereum main chain. However, the economic value of the application tokens need to be maintained, thus the system should allow purchasing the tokens on the main chain, paying with actual ethers for acquiring decentralized applications specific tokens, and then moving the tokens on a separate chain to use them in the actual application. Plasma [62] is an alternative proposal of the sidechains on Ethereum. An equivalent transaction is generated on the sidechain (plasma chain) from nothing, giving the corresponding coins to the plasma chain user. The validation mechanism is based on Fraud Proofs by periodically checking the main chain. If a user wants to spend its coins on the main chain in a fraudulent manner, its action can be proved to be malicious by submitting proof of a spending transaction that was previously registered on the plasma chain. This would prevent the main chain transaction from being validated. Since a user can issue a spending transaction directly from the main chain this offers a great advantage over the classical sidechain approach. That is, in case the plasma chain is compromised, the users can still issue their withdrawals. Moreover, Plasma is designed to permit the implementation of nested chains, thus creating a tree-structured system of chains that requires each user to monitor only the chains that can affect one's transactions.

Table 8 presents the main solutions for inter blockchain ledgers interoperability and communication comparing their main features. While hybrid models and Plasma may offer reliable solutions in terms of ledger-to-ledger interaction, the Lightning Network can be considered a viable alternative that can provide high transfer rates.

Table 8. Ledger to Ledger interoperability approaches.

	Implementation Level	Deposit Mechanism	Withdraw Mechanism	Trusted Entities	Chain Independence
Central Exchange	Escrow	TX to a central authority	TX to a central authority	Central Authority	Yes
MultiSig Federation [138]	Escrow	TX to a multi-signature federation	TX to a multi-signature federation	N Delegates	Yes
Entangled Chain [140]	Protocol Layer	SPV Proof from the main chain	-	Dependent on the withdraw	Mining rate restriction
Sidechains [121,122]	Network & Protocol Layer	SPV Proof from the main chain + proof of block validity	SPV Proof from the sidechain + proof of block validity	Sidechain miners	Yes
Drivechain [142]	Network & Protocol Layer	SPV Proof from the main chain + proof of block validity	SPV Proof from the sidechain + proof of block validity + miners votes	Sidechain miners	Yes
Hybrid Models [123]	Network & Protocol Layer	SPV Proof from the main chain + proof of block validity	SPV Proof from the sidechain + proof of block validity + miners votes + multi-signature notaries	Sidechain miners + Notaries	Yes
Plasma [62]	Network & Protocol Layer	Proof for TX on the main chain	Direct withdraw + Fault Proofs	Central Authority, N Delegates	Yes
Lightning Network, Interledger [137]	Scalability Tier Solution	Atomic Swap	Atomic Swap	None	Yes

5. Discussion and Development Guidelines

As presented in the previous sections there are a lot of distributed ledger technologies. Most of them have emerged in the last years to address specific development issues. The ecosystem is subject to rapid changes making the selection of technologies for the implementation of decentralized applications rather difficult and fuzzy. At the same time DLTs bring benefits concerning the implementation of decentralized applications. They eliminate the need for a mediator, having the capabilities to enforce contract rules on-chain, each participant being aware of the consequences of his actions. The hashed data structures allow for easy traceability of the assets and state updates in the ledger. Since the records are public and replicated, great transparency is provided. Even if all the transactions and all the actions are public the platforms provide high security through consensus, public-key cryptography, and tamper-resistant recording. Nevertheless, a lot of nowadays applications and management systems are rather centralized (e.g., utility grids, banking, stock exchange, etc.). Thus, efforts are committed to investigating how the DLT can be applied, integrated, and used for decentralizing such applications and systems.

In the rest of this section, we present a guideline for decentralizing systems by designing and implementing decentralized applications.

5.1. Decentralization of Design

The main steps required for decentralizing a centralized system by designing decentralized applications are detailed below discussing their role and specific technological requirements (see Figure 8).

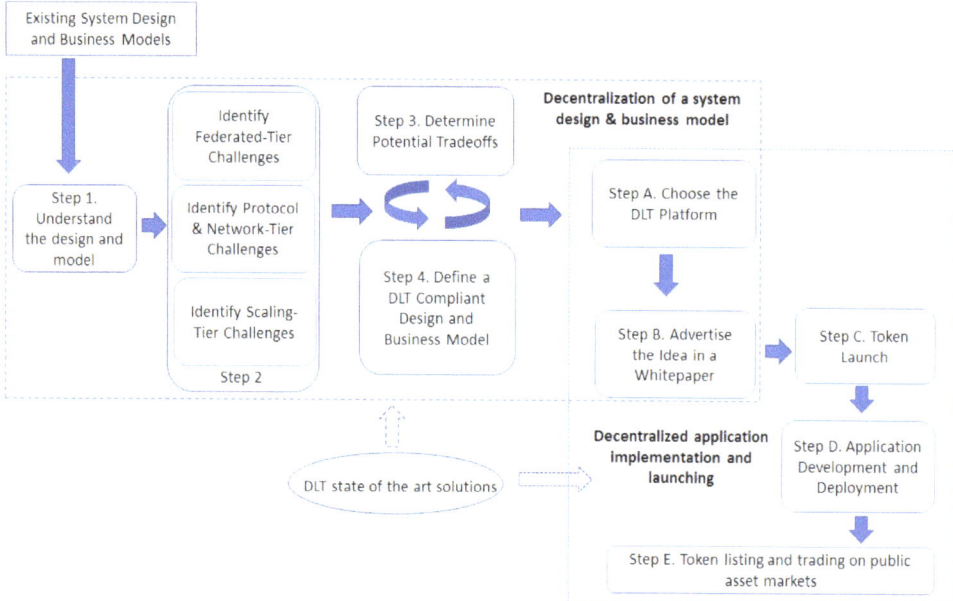

Figure 8. Decentralized application design and implementation steps.

Step 1. *Understanding the existing system design and business model.* An exhaustive understanding of the business is required to design a decentralized application. Good knowledge of the business model and rules that govern the domain, as well as of the architecture used in the centralized implementation, are prerequisites for starting the sequence of steps required to achieve a consistent and well-documented decentralized solution. One needs to identify the functional and non-functional requirements that need to be addressed for ensuring the completeness of the design. Most of the time, a complete and sound list of requirements can be obtained by holding several interview sessions with the future stakeholders of the system.

Step 2. *Identify potential challenges for each tier.* Once the functional and non-functional requirements are clearly defined, the challenges that can arise once mapping the model on a decentralized solution can already be identified. We classify the challenges based on the tiers identified at the level of architecture. Protocol and Network Tier challenges: What is the targeted network level, public or private? Does the system need to hide the transacted information? Is selective disclosure a requirement of the business? Etc. Scalability Tier challenges: Are the scalability requirements higher than what the Protocol and Network Tier solutions can offer? Are the scalability requirements targeting storage, throughput, or computation?

Step 3. *Determine potential tradeoffs.* Depending on the set of challenges identified, there might be situations where not all the requirements can be successfully ensured by the DLT solutions. In such situations, a tradeoff needs to be made. Most of the time, the application scalability may be affected while incorporating the reliability, immutability, and consensus properties ensured by a DLT. Furthermore, by outsourcing components of the system to the Scalability Tier, another tradeoff regarding the actual decentralization must be done since most of the Scalability Tier solution requires a trusted party to oversee the outsourced component.

Step 4. *Define a DLT compliant translation of the application design and business model.* It is important to conduct thorough research of the state of the art whenever a DLT solution is considered, for a system decentralization. Obtaining a clear picture of all potential theologies alternatives to the problems identified can be quite difficult. A decentralized design of the system should be proposed, and the most suitable solution for each architectural component needs to be selected.

For example, if smart contracts technology is selected for implementing the application business logic several key features need to be carefully investigated. The smart contracts can keep track of the data stored and act as a financial escrow for the interacting parties. The functions of a smart contract can be used internally by the contract, or can be exposed as an API to the external modules. Most of the time the smart contracts are used as state machines, where the state is updated according to the latest input received. To interrogate a DLT event-based information is considered. The events are elements emitted by the contracts during processing and stored in tamper-resistant structures as well. Instructions executed by a smart contract in a public chain have a processing cost paid by the transaction issuer. The payment is directed to the miner, for using its computational resources. Consequently, an appropriate data structure must always be used to avoid high costs. The processing rules should be implemented, if possible, in the same contract where the data is stored, to avoid the cost of referencing other contracts. When deploying a contract, the storage of the code is also paid by the issuer. For multiple deployments of the same contract, a good approach is to use libraries containing static code that is deployed only once and then linked to each of the deployed contract instances.

From this point forward, a classical pipeline of implementation, verification, and maintenance can be applied to validate the final application.

5.2. Decentralized Application Development

Upon successful implementation of the above-presented steps, a decentralized design of the system and a DLT compliant translation of the business model is obtained. There are several steps to go through for launching such a decentralized application, from DLT platform selection, advertising, and crowdfunding up to the actual development and operationalization (see Figure 8).

Step A. *Choose the DLT platform.* From this point forward one needs to proceed with the evaluation of the existing DLT implementation platforms, since some decentralized applications may be compatible with existing systems, thus beneficiating from the already built network of miners. Two cases may emerge (Table 9): an existing platform matches the decentralized application requirements, or a new custom chain needs to be implemented. In the latter case, a custom chain is built and used as the base PN-Tier for the decentralized application implementation. Most of the time, this situation arises when the current frameworks' specifications are not compliant with the application requirements, thus a new DLT core platform must be developed, and a new network of nodes must be built. Furthermore, a tradeoff regarding the cost and the complexity of the implementation must be considered when choosing one of the two solutions.

Step B. *Publish and advertise the idea through a whitepaper.* One principle of decentralized application whitepaper is to provide complete transparency to build trust with the investors. Companies are encouraged to provide an honest technical and economic roadmap. They should explain in detail the technical feasibility of the solution as well as the investment and revenue plans, the shares among the company partners, the economic sustainability of the solution, etc.

Step C. *Token launch or Initial Coin Offering.* A fixed number of tokens should be released to attract investors and raise money for the development phase. The token distribution plan specifies the maximum number of tokens that will be ever generated by the decentralized application, the tokens unlocked for distribution, the tokens transferred to the founders or other participants, and the tokens locked for future use. The initial coin offering plans are advertised by the company, mentioning the initial price, start date, end date, the number of tokens unlocked, pricing schemes (constant pricing, incremental pricing, etc.). The token registry and the distribution rules are programmed as smart contracts and deployed on-chain. Each buyer willing to invest in the decentralized application will need to acquire the necessary cryptocurrency and sign a transaction paying the requested sum in return for several tokens. The distribution contract will validate the deposited sum, and if all the rules hold, the token registry will update the buyer's account accordingly.

Step D. *Development and application deployment.* The decentralized application development will continue according to the roadmap presented in the whitepaper. Based on the alternatives presented in the previous chapters, a guideline for choosing the specific implementation solutions is presented in Table 10. Upon finalization, the application will be deployed and all the users that acquired tokens during the initial coin offering will be able to start using the decentralized application or will be able to sell the tokens to other interested parties.

Step E. *Token listing on public exchanges.* To facilitate the exchange of tokens between interested users, the company can list the token on one of the public exchanges. Sellers can make their offers and buyers can make their bids therefore, depending on the public interest in the launched business, the token can have the potential to raise its value. Being an off-chain exchange, only upon settlement the updated balance will be registered on-chain mirroring the transaction that happened between the seller and the buyer. However, some of these exchanges chose to act as a custodian for the exchanged tokens and the actual registry is not updated on-chain.

Table 9. DLT platform choice.

Option	Advantages	Disadvantages
Standalone Customized DLT	Target the business requirements, limitations, and challenges	Needs to build a network and gain trust in the mining nodes. Increased implementation complexity
Existing DLT Platform	Existing P2P Network and community Attacks are unlikely considering the high number of existing nodes	High costs imposed by the mining nodes Some tradeoffs may be made due to the core-properties of an existing DLT

Table 10. DLT technologies and alternatives.

Issue	Tier	Description	Alternative Solutions
Asset Representation	PN-Tier	Custom tokens in an existing DLT	**ERC-721**: CryptoKitties [23], Rarible [22], EtherTulips [20] **ERC20**: Grid [21], Telcoin [25], Storj [110]
		Native token in a custom build DLT	Filecoin [112], MediaChain [113]
Data Structure	PN-Tier	Selection of Security and Decentralisation over Scalability Selection of Scalability and Security over Decentralization	**Blockchain platforms**: Ethereum [15], Hyperledger [34], Quorum [49] **Directed Acyclic Graphs**: IOTA [16], HashGraph [39]
Data privacy	PN-Tier	Anonymity of identities involved	**Ring Signatures**: Monero [18], CryptoNote [54] **Mixers**: CoinJoin [47], Dash [48]
		Anonymity of content retaining verifiability	Homomorphic Encryptions (MimbleWimble [53]); **Secure Multi Party Computation**: Enigma [133]; **Zero-Knowledge Proofs**: Zcash [19], Zerocoin [51],
	PN-Tier/ S-Tier	Anonymity of data using references to locations in Scalability Tier	**External Storage System**: IPFS [111]

Table 10. Cont.

Issue	Tier	Description	Alternative Solutions
Business Enforcement	PN-Tier	Built-in custom business rules	Filecoin [112], NRGCoin [24]
		Generic rules and pluggable Turing Complete implementation	Ethereum [15], Hyperledger [34]
Data Propagation	PN-Tier	Selective disclosure	Hyperledger Multichannel (Corda [61], Plasma [62]), Quorum [49]
		Global disclosure	Any public permissionless DLT
Permissions & Consensus	PN-Tier	Public Permissionless & Byzantine Fault Tolerance	Ethereum with PoS or PoW
		Public Permissioned	Ethereum PoS or PoA; Hyperledger [34]
		Consortium or Private	Hyperledger [34], Corda [61]
Large Size data	PN-Tier	Selection of Scalability and Decentralisation over Security	BigChainDB [20]
	PN-Tier/ S-Tier	File System Storage	IPFS [111], Storj [110], Filecoin [112]
High-Frequency Data	PN-Tier/S-Tier	Selection of Scalability and Security over Decentralization	IOTA [16], Sharding [104], Sidechains [121]
		Overlay Networks; Payment/State Channels	Payment Channels [124], Raiden [125]
Computational Intensive Algorithms	PN-Tier/S-Tier	Selection of Scalability and Security over Decentralization	**Secure Multi-Party Computation**: Enigma [133]; TrueBit [129]
		Oracles	Provable [128]

6. Conclusions

The distributed ledger technology has the potential of being a game-changer in many domains, its recent developments being triggered not only by technology expectations but also by social ones. DLTs have enabled the development of decentralized applications but this process is still complicated due to the high availability of new and rather immature technological solutions that address different implementation issues. DLTs and especially blockchain are an effervescent innovation area. Therefore, the review of technological solutions and implementation guidelines are needed to improve understanding and to ease the development of decentralized applications.

Most of the similar initiatives found in the literature are focusing on the blockchain-based distributed ledger and pay little attention to other DLT implementations. In this paper, we have provided a review of the existing DLT solutions. We highlight their applicability, advantages, and disadvantages concerning other technologies. A significant number of solutions have been reviewed to provide a holistic image of the DLT implementation variations. In the technology review process, we have considered references from both academia and the private sector.

At the same time, our review was driven by a decentralized application architecture that has been used and validated in previous publications. The DLT solutions are categorized according to the 3-tiers of the architecture thus the criteria for selecting the technologies and organizing the review are strictly related to the development issues that may be encountered and need to be addressed in each tier.

Finally, we provide a guideline for decentralized application development defining specific steps for decentralizing a system design and business model using DLT and for implementing and launching it as a decentralized application. No similar guideline could be found in the reviewed literature. The guideline can be used to streamline the nowadays efforts for investigating how the DLT can be applied, integrated, and used for decentralizing systems such as medical systems, electricity grids, financial sector, etc., and for implementing new decentralized applications.

Author Contributions: Conceptualization, T.C. and I.S.; methodology, T.C. and I.A.; investigation, C.A. and M.A.; writing—original draft preparation, C.A. and M.A.; writing—review and editing, I.S. and T.C.; visualization, I.A.; supervision, I.S. All authors have read and agreed to the published version of the manuscript.

Funding: This research was funded by the European Union's Horizon 2020 research and innovation program grant number 957816 (BRIGHT) and grant number 774478 (eDREAM).

Data Availability Statement: Not Applicable, the study does not report any data.

Conflicts of Interest: The authors declare no conflict of interest. The funders had no role in the design of the study; in the collection, analyses, or interpretation of data; in the writing of the manuscript, or in the decision to publish the results.

References

1. Top Trends in the Gartner Hype Cycle for Emerging Technologies. 2016. Available online: https://www.gartner.com/en/newsroom/press-releases/2016-08-16-gartners-2016-hype-cycle-for-emerging-technologies-identifies-three-key-trends-that-organizations-must-track-to-gain-competitive-advantage (accessed on 24 February 2021).
2. Top Trends in the Gartner Hype Cycle for Emerging Technologies. 2018. Available online: https://www.gartner.com/smarterwithgartner/5-trends-emerge-in-gartner-hype-cycle-for-emerging-technologies-2018/ (accessed on 24 February 2021).
3. Top Trends in the Gartner Hype Cycle for Emerging Technologies. 2019. Available online: https://www.gartner.com/smarterwithgartner/5-trends-appear-on-the-gartner-hype-cycle-for-emerging-technologies-2019/ (accessed on 24 February 2021).
4. Nakamoto, S. Bitcoin: A Peer-To-Peer Electronic Cash System. 2008. Available online: https://bitcoin.org/bitcoin.pdf (accessed on 24 February 2021).
5. Pop, C.; Antal, M.; Cioara, T.; Anghel, I.; Sera, D.; Salomie, I.; Raveduto, G.; Ziu, D.; Croce, V.; Bertoncini, M. Blockchain-Based Scalable and Tamper-Evident Solution for Registering Energy Data. *Sensors* **2019**, *19*, 3033. [CrossRef]
6. Pop, C.; Cioara, T.; Antal, M.; Anghel, I.; Salomie, I.; Bertoncini, M. Blockchain Based Decentralized Management of Demand Response Programs in Smart Energy Grids. *Sensors* **2018**, *18*, 162. [CrossRef] [PubMed]
7. Farahani, B.; Firouzi, F.; Luecking, M. The convergence of IoT and distributed ledger technologies (DLT): Opportunities, challenges, and solutions. *J. Netw. Comput. Appl.* **2021**, *177*, 102936. [CrossRef]
8. Maesa, D.; Mori, P. Blockchain 3.0 applications survey. *J. Parallel Distrib. Comput.* **2020**, *138*, 99–114. [CrossRef]
9. FBenčić, M.; Žarko, I.P. Distributed Ledger Technology: Blockchain Compared to Directed Acyclic Graph. In Proceedings of the 2018 IEEE 38th International Conference on Distributed Computing Systems (ICDCS), Vienna, Austria, 2–6 July 2018; pp. 1569–1570. [CrossRef]
10. Zia, M.F.; Benbouzid, M.; Elbouchikhi, E.; Muyeen, S.M.; Techato, K.; Guerrero, J.M. Microgrid Transactive Energy: Review, Architectures, Distributed Ledger Technologies, and Market Analysis. *IEEE Access* **2020**, *8*, 19410–19432. [CrossRef]
11. Pop, C.; Antal, M.; Cioara, T.; Anghel, I. Trading Energy as a Digital Asset: A Blockchain based Energy Market. In *Cryptocurrencies and Blockchain Technologies and Applications: Decentralization and Smart Contracts*; Gulshan, S., Nhuong, L., Kavita, S., Eds.; Wiley-Scrivener: Hoboken, NJ, USA, 2020; ISBN 978-1-119-62116-4.
12. Pop, C.D.; Antal, M.; Cioara, T.; Anghel, I.; Salomie, I. Blockchain and Demand Response: Zero-Knowledge Proofs for Energy Transactions Privacy. *Sensors* **2020**, *20*, 5678. [CrossRef]
13. Pop, C.; Pop, C.; Marcel, A.; Vesa, A.; Petrican, T.; Cioara, T.; Anghel, I.; Salomie, I. Decentralizing the Stock Exchange using Blockchain An Ethereum-based implementation of the Bucharest Stock Exchange. In Proceedings of the 2018 IEEE 14th International Conference on Intelligent Computer Communication and Processing (ICCP), Cluj-Napoca, Romania, 6–8 September 2018; pp. 45–466. [CrossRef]
14. Pop, C.D.A.; Cioara, T.; Antal, M.; Anghel, I. Blockchain Platform for COVID-19 Vaccine Supply Management. 2020. Available online: https://arxiv.org/abs/2101.00983 (accessed on 24 February 2021).
15. Wood, G. Ethereum: A Secure Decentralised Generalised Transaction Ledger. Ethereum Project Yellow Paper 151.2014 (2014): 1–32. Available online: http://gavwood.com/paper.pdf (accessed on 24 February 2021).
16. Pervez, H.; Muneeb, M.; Irfan, M.U.; Haq, I.U. A Comparative Analysis of DAG-Based Blockchain Architectures. In Proceedings of the 12th International Conference on Open Source Systems and Technologies (ICOSST), Lahore, Pakistan, 19–21 December 2018; pp. 27–34. [CrossRef]
17. Branson, E. *Litecoin: The Ultimate Beginner's Guide for Understanding Litecoins and What You Need to Know*; CreateSpace Independent Publishing Platform: Scotts Valley, CA, USA, 2014; ISBN 1507878192.
18. Alonso, K.M. Zero to Monero. Available online: https://www.getmonero.org/library/Zero-to-Monero-1-0-0.pdf (accessed on 24 February 2021).
19. Sasson, E.B.; Chiesa, A.; Garman, C.; Green, M.; Miers, I.; Tromer, E.; Virza, M. Zerocash: Decentralized anonymous payments from bitcoin. In Proceedings of the 2014 IEEE Symposium on Security and Privacy, Berkeley, CA, USA, 18–21 May 2014. [CrossRef]
20. EtherTulips. Available online: https://ethertulips.com/ (accessed on 24 February 2021).

21. Grid. Available online: https://web.gridplus.io/grid-token (accessed on 24 February 2021).
22. Rarible. Available online: https://rarible.com/ (accessed on 24 February 2021).
23. CryptoKitties: Collectible and Breedable Cats Empowered by Blockchain Technology. Available online: http://upyun-assets.ethfans.org/uploads/doc/file/25583a966d374e30a24262dc5b4c45cd.pdf?_upd=CryptoKitties_WhitePapurr_V2.pdf (accessed on 24 February 2021).
24. NRGcoin. Available online: https://nrgcoin.org/ (accessed on 24 February 2021).
25. TelCoin. Available online: https://www.telco.in/ (accessed on 24 February 2021).
26. Ethereum Improvement Proposals. Available online: http://eips.ethereum.org/erc (accessed on 24 February 2021).
27. Blocksquare. Available online: https://blocksquare.io/ (accessed on 24 February 2021).
28. Blockchain in Commercial Real Estate: The Future Is Here. Available online: https://www2.deloitte.com/us/en/pages/financial-services/articles/blockchain-in-commercial-real-estate.html (accessed on 24 February 2021).
29. Zeilinger, M. Digital art as 'onetised graphics': Enforcing intellectual property on the blockchain. *Philos. Technol.* **2018**, *31*, 15–41. [CrossRef]
30. Nielson, B. Blockchain Ownership of Intellectual Property. Available online: http://www.yourtrainingedge.com/blockchain-ownership-of-intellectual-property/ (accessed on 24 February 2021).
31. Turkanovi, M.; Hölbl, M.; Koši, K.; Heriko, M.; Kamišali, A. EduCTX: A blockchain-based higher education credit platform. *IEEE Access* **2018**, *6*, 5112–5127. [CrossRef]
32. Durant, E.; Trachy, A. Digital Diploma Debuts at MIT. Available online: http://news.mit.edu/2017/mit-debuts-secure-digital-diploma-using-bitcoin-blockchain-technology-1017 (accessed on 24 February 2021).
33. del Castillo, M. Britain's Royal Mint Reveals Details on "Live" Blockchain for Tracking Gold. Available online: https://www.coindesk.com/britains-royal-mint-reveals-details-on-live-blockchain-for-tracking-gold/ (accessed on 24 February 2021).
34. Androulaki, E.; Barger, A.; Bortnikov, V.; Cachin, C.; Christidis, K.; De Caro, A. Hyperledger fabric: A distributed operating system for permissioned blockchains. In Proceedings of the Thirteenth EuroSys Conference, Porto, Portugal, 23–26 April 2018. [CrossRef]
35. Nicolas van Saberhagen, CryptoNode v 2.0, Monero White Paper. 2013. Available online: https://bytecoin.org/old/whitepaper.pdf (accessed on 24 February 2021).
36. David, S.; Youngs, N.; Britto, A. The Ripple Protocol Consensus Algorithm. Ripple Labs Inc. White Paper 5 2014. Available online: https://ripple.com/files/ripple_consensus_whitepaper.pdf (accessed on 24 February 2021).
37. Churyumov, A. Byteball: A Decentralized System for Storage and Transfer of Value. Available online: https://byteball.org/Byteball.pdf (accessed on 24 February 2021).
38. Dagcoin Whitepaper. Available online: https://dagcoin.org/whitepaper/ (accessed on 24 February 2021).
39. Baird, L. The Swirlds Hashgraph Consensus Algorithm: Fair, Fast, Byzantine Fault Tolerance. Available online: https://www.swirlds.com/downloads/SWIRLDS-TR-2016-01.pdf (accessed on 24 February 2021).
40. Braun, E.H.; Luck, N.; Brock, A. Holochain-Scalable Agent-Centric Distributed Computing. 2018. Available online: https://github.com/holochain/holochain-proto/blob/whitepaper/holochain.pdf (accessed on 24 February 2021).
41. Chen, J. Flowchain: A Distributed Ledger Designed for Peer-to-Peer IoT Networks and Real-Time Data Transactions. In Proceedings of the 2nd International Workshop on Linked Data and Distributed Ledgers, Portoroz, Slovenia, 29 May 2017.
42. The Coordicide. 2019. Available online: https://cdn0.tnwcdn.com/wp-content/blogs.dir/1/files/2019/05/Coordicide_WP.pdf (accessed on 24 February 2021).
43. Kosba, A.; Miller, A.; Shi, E.; Wen, Z.; Papamanthou, C. Hawk: The blockchain model of cryptography and privacy-preserving smart contracts. In Proceedings of the 2016 IEEE symposium on security and privacy (SP), San Jose, CA, USA, 22–26 May 2016. [CrossRef]
44. Azaria, A.; Ekblaw, A.; Vieira, T.; Lippman, A. Medrec: Using blockchain for medical data access and permission management. In Proceedings of the 2016 2nd International Conference on Open and Big Data (OBD), Vienna, Austria, 22–24 August 2016; pp. 25–30. [CrossRef]
45. Griggs, K.N.; Ossipova, O.; Kohlios, C.P.; Baccarini, A.N.; Howson, E.A.; Hayajneh, T. Healthcare blockchain system using smart contracts for secure automated remote patient monitoring. *J. Med Syst.* **2018**, *42*, 130. [CrossRef]
46. Elagin, V.; Spirkina, A.; Levakov, A.; Belozertsev, I. Blockchain Behavioral Traffic Model as a Tool to Influence Service IT Security. *Future Internet* **2020**, *12*, 68. [CrossRef]
47. Maurer, F.K.; Neudecker, T.; Florian, M. Anonymous CoinJoin transactions with arbitrary values. In Proceedings of the 2017 IEEE Trustcom/BigDataSE/ICESS, Sydney, Australia, 1–4 August 2017; pp. 522–529. [CrossRef]
48. Duffield, E.; Diaz, D. Dash: A Payments-Focused Cryptocurrency. 2018. Available online: https://github.com/dashpay/dash/wiki/Whitepaper (accessed on 24 February 2021).
49. Quorum Whitepaper. Available online: https://github.com/jpmorganchase/quorum-docs/blob/master/Quorum%20Whitepaper%20v0.1.pdf (accessed on 24 February 2021).
50. Goldreich, O.; Micali, S.; Wigderson, A. Proofs that yield nothing but the validity of their assertion. *Preprint* **1986**.
51. Miers, I.; Garman, C.; Green, M.; Rubin, A.D. Zerocoin: Anonymous distributed e-cash from bitcoin. In Proceedings of the 2013 IEEE Symposium on Security and Privacy, Berkeley, CA, USA, 19–22 May 2013. [CrossRef]

52. Ben-Sasson, E.; Chiesa, A.; Tromer, E.; Virza, M. Succinct non-interactive zero knowledge for a von Neumann architecture. In Proceedings of the 23rd {USENIX} Security Symposium ({USENIX} Security 14, San Diego, CA, USA, 20–22 August 2014; pp. 781–796.
53. Poelstra, A. Mimblewimble. Self-Published in October 2016. Available online: https://download.wpsoftware.net/bitcoin/wizardry/mimblewimble.pdf (accessed on 24 February 2021).
54. van Saberhagen, N. CryptoNode v 2.0, Monero White Paper. 2016. Available online: https://github.com/monero-project/research-lab/blob/master/whitepaper/whitepaper.pdf (accessed on 24 February 2021).
55. Garrick, H.; Rauchs, M. 2017 Global Blockchain Benchmarking Study. Available online: https://papers.ssrn.com/sol3/papers.cfm?abstract_id=3040224 (accessed on 24 February 2021).
56. Morrison, D.R. PATRICIA-practical algorithm to retrieve information coded in alphanumeric. *J. ACM* **1968**, *15*, 514–534. [CrossRef]
57. Nxt. Available online: https://nxtdocs.jelurida.com/Nxt_Whitepaper (accessed on 24 February 2021).
58. Counterparty. Available online: https://counterparty.io/docs/ (accessed on 24 February 2021).
59. Saia, R.; Carta, S.; Recupero, D.; Fenu, G. Internet of Entities (IoE): A Blockchain-based Distributed Paradigm for Data Exchange between Wireless-based Devices. In Proceedings of the 8th International Conference on Sensor Networks (SENSORNETS 2019), Prague, Czech Republic, 26–27 February 2019; pp. 77–84. [CrossRef]
60. Honar Pajooh, H.; Rashid, M.; Alam, F.; Demidenko, S. Hyperledger Fabric Blockchain for Securing the Edge Internet of Things. *Sensors* **2021**, *21*, 359. [CrossRef] [PubMed]
61. Palm, E.; Bodin, U.; Schelén, O. Approaching Non-Disruptive Distributed Ledger Technologies via the Exchange Network Architecture. *IEEE Access* **2020**, *8*, 12379–12393. [CrossRef]
62. Joseph, P.; Buterin, V. Plasma: Scalable Autonomous Smart Contracts. White Paper. 2017, pp. 1–47. Available online: https://plasma.io/plasma.pdf (accessed on 24 February 2021).
63. George, C.; Dollimore, J.; Kindberg, T. *Distributed Systems: Concepts and Design*, 3rd ed.; Addison-Wesley: Boston, MA, USA, 2001; p. 452, ISBN 978-0201-61918-8.
64. Huang, D.; Ma, X.; Zhang, S. Performance Analysis of the Raft Consensus Algorithm for Private Blockchains. *IEEE Trans. Syst. ManCybern. Syst.* **2020**, *50*, 172–181. [CrossRef]
65. Lamport, L. *The Part-Time Parliament, ACM Transactions on Computer Systems 16*. 1998. Available online: https://lamport.azurewebsites.net/pubs/lamport-paxos.pdf (accessed on 24 February 2021).
66. Diego, O.; Ousterhout, J. In search of an understandable consensus algorithm. In Proceedings of the 2014 {USENIX} Annual Technical Conference, Philadelphia, PA, USA, 19–20 June 2014.
67. Longo, R.; Podda, A.S.; Saia, R. Analysis of a Consensus Protocol for Extending Consistent Subchains on the Bitcoin Blockchain. *Computation* **2020**, *8*, 67. [CrossRef]
68. Pires, M.; Ravi, S.; Rodrigues, R. Generalized Paxos Made Byzantine (and Less Complex). *Algorithms* **2018**, *11*, 141. [CrossRef]
69. Castro, M.; Liskov, B. Practical Byzantine Fault Tolerance and Proactive Recovery. *ACM Trans. Comput. Syst.* **2002**, *20*, 398–461. [CrossRef]
70. Clement, A.; Wong, E.; Alvisi, L.; Dahlin, M.; Marchetti, M. Making Byzantine Fault Tolerant Systems Tolerate Byzantine Faults. In *Networked Systems Design and Implementation*; USENIX: Berkeley, CA, USA, 2009.
71. Aublin, P.-L.; Mokhtar, S.B.; Quéma, V. RBFT: Redundant Byzantine Fault Tolerance. In Proceedings of the 33rd IEEE International Conference on Distributed Computing Systems, Philadelphia, PA, USA, 8–11 July 2013. [CrossRef]
72. Abd-El-Malek, M.; Ganger, G.; Goodson, G.; Reiter, M.; Wylie, J. Fault-scalable Byzantine Fault-Tolerant Services. *ACM Sigops Oper. Syst. Rev.* **2005**, *39*, 59. [CrossRef]
73. Cowling, J.; Myers, D.; Liskov, B.; Rodrigues, R.; Shrira, L. HQ Replication: A Hybrid Quorum Protocol for Byzantine Fault Tolerance. In Proceedings of the 7th USENIX Symposium on Operating Systems Design and Implementation, Seattle, WA, USA, 6–8 November 2006; pp. 177–190, ISBN 1-931971-47-1.
74. Kotla, R.; Alvisi, L.; Dahlin, M.; Clement, A.; Wong, E. Zyzzyva: Speculative Byzantine Fault Tolerance. *ACM Trans. Comput. Syst.* **2009**, *27*, 1–39. [CrossRef]
75. Guerraoui, R.; Kneževic, N.; Vukolic, M.; Quéma, V. The Next 700 BFT Protocols. In Proceedings of the 5th European conference on Computer systems, Paris, France, 30 March–2 April 2010. [CrossRef]
76. Zhao, W. A Byzantine Fault Tolerant Distributed Commit Protocol. In Proceedings of the Third IEEE International Symposium on Dependable, Autonomic and Secure Computing (DASC 2007), Columbia, MD, USA, 25–26 September 2007; pp. 37–46. [CrossRef]
77. Markus, J.; Ari, J. *Proofs of Work and Bread Pudding Protocols, Communications and Multimedia Security*. 1999. Available online: http://www.hashcash.org/papers/bread-pudding.pdf (accessed on 24 February 2021).
78. Goldwasser, S.; Micali, S.; Rackoff, C. The Knowledge Complexity of Interactive Proof-Systems. Available online: http://citeseerx.ist.psu.edu/viewdoc/download?doi=10.1.1.419.8132&rep=rep1&type=pdf (accessed on 24 February 2021).
79. Sompolinsky, Y.; Zohar, A. Secure High-Rate Transaction Processing in Bitcoin. In *Financial Cryptography and Data Security*; Böhme, R., Okamoto, T., Eds.; Springer: Berlin/Heidelberg, Germany, 2015.
80. Cocco, L.; Pinna, A.; Marchesi, M. Banking on Blockchain: Costs Savings Thanks to the Blockchain Technology. *Future Internet* **2017**, *9*, 25. [CrossRef]
81. Kim, H.; Kim, K.; Kwon, H.; Seo, H. ASIC-Resistant Proof of Work Based on Power Analysis of Low-End Microcontrollers. *Mathematics* **2020**, *8*, 1343. [CrossRef]

82. Franco, P. *Understanding Bitcoin: Cryptography, Engineering and Economics*; Wiley: Hoboken, NJ, USA, 2014; ISBN 978-1-119-01916-9.
83. Vujičić, D.; Jagodić, D.; Ranđić, S. Blockchain technology, bitcoin, and Ethereum: A brief overview. In Proceedings of the 2018 17th International Symposium INFOTEH-JAHORINA (INFOTEH), East Sarajevo, Bosnia and Herzegovina, 21–23 March 2018; pp. 1–6. [CrossRef]
84. Biryukov, A.; Khovratovich, D. Equihash: Asymmetric Proof-of-Work Based on the Generalized Birthday Problem, Network and Distributed System Security Symposium. 2016. Available online: https://www.ndss-symposium.org/wp-content/uploads/2017/09/equihash-asymmetric-proof-of-work-based-generalized-birthday-problem.pdf (accessed on 24 February 2021).
85. Tromp, J. Cuckoo Cycle: A memory bound graph-theoretic proof-of-work. In *International Conference on Financial Cryptography and Data Security*; Springer: Berlin/Heidelberg, Germany, 2015; pp. 49–62.
86. GRIN. Available online: https://github.com/ignopeverell/grin (accessed on 24 February 2021).
87. AEternity. Available online: http://www.aeternity.com/ (accessed on 24 February 2021).
88. Yang, Z.; Yang, K.; Lei, L.; Zheng, K.; Leung, V.C.M. Blockchain-Based Decentralized Trust Management in Vehicular Networks. *IEEE Internet Things J.* **2018**, *6*, 1495–1505. [CrossRef]
89. CureCoin. Available online: https://www.curecoin.net/ (accessed on 24 February 2021).
90. Bentov, I.; Lee, C.; Mizrahi, A.; Rosenfeld, M. Proof of Activity: Extending Bitcoin's Proof of Work via Proof of Stake. *Acm Sigmetrics Perform. Eval. Rev.* **2014**, *42*, 34–37. [CrossRef]
91. Christina, J. DTB001: Decred Technical Brief. Available online: https://cryptorating.eu/whitepapers/Decred/decred.pdf (accessed on 24 February 2021).
92. Boni, K.R.C.; Xu, L.; Chen, Z.; Baddoo, T.D. A Security Concept Based on Scaler Distribution of a Novel Intrusion Detection Device for Wireless Sensor Networks in a Smart Environment. *Sensors* **2020**, *20*, 4717. [CrossRef] [PubMed]
93. Buterin, V.; Griffith, V. Casper the friendly finality gadget. *arXiv* **2017**, arXiv:1710.09437. Available online: https://arxiv.org/abs/1710.09437 (accessed on 24 February 2021).
94. Chen, J.; Micali, S. Algorand: A secure and efficient distributed ledger. *Theor. Comput. Sci.* **2019**. [CrossRef]
95. Buterin, V. Slasher: A Punitive Proof-of-Stake Algorithm. 2014. Available online: https://blog.ethereum.org/2014/01/15/slasher-a-punitive-proof-of-stake-algorithm/ (accessed on 24 February 2021).
96. Bentov, I. Cryptocurrencies without Proof of Work. 2017. Available online: https://fc16.ifca.ai/bitcoin/papers/BGM16.pdf (accessed on 24 February 2021).
97. Zhu, S.; Cai, Z.; Hu, H.; Li, Y.; Li, W. zkCrowd: A Hybrid Blockchain-Based Crowdsourcing Platform. *IEEE Trans. Ind. Inform.* **2020**, *16*, 4196–4205. [CrossRef]
98. Elrond. Available online: https://elrond.com/ (accessed on 24 February 2021).
99. Maofan, Y.; Dahlia, M.; Michael, R.; Guy, G.; Ittai, A. HotStuff: BFT Consensus with Linearity and Responsiveness. In Proceedings of the 2019 ACM Symposium on Principles of Distributed Computing, Toronto, ON, Canada, 29 July–2 August 2019; pp. 347–356. [CrossRef]
100. Kuhn, R.; Yaga, D.; Voas, J. Rethinking Distributed Ledger Technology. *Computer* **2019**, *52*, 68–72. [CrossRef]
101. Bitcoin: Maximum Transactions Rate. Available online: https://en.bitcoin.it/wiki/Maximum_transaction_rate (accessed on 21 January 2021).
102. Ehrsam, F. Scalability Ethereum to Billions of Users. Available online: https://medium.com/@FEhrsam/scalability-ethereum-to-billions-of-users-f37d9f487db1 (accessed on 24 February 2021).
103. Costa, C.H.; Vianney, B.M.; Filho, J.; Henrique, M.; Maia, P.; Carlos, M.B.; Oliveira, F. Sharding by Hash Partitioning—A Database Scalability Pattern to Achieve Evenly Sharded Database Clusters. In Proceedings of the 17th International Conference on Enterprise Information Systems, Barcelona, Spain, 27–30 April 2015; pp. 313–320, ISBN 978-989-758-096-3. [CrossRef]
104. Yu, G.; Wang, X.; Yu, K.; Ni, W.; Zhang, J.A.; Liu, R.P. Survey: Sharding in Blockchains. *IEEE Access* **2020**, *8*, 14155–14181. [CrossRef]
105. Chow, S.S.M.; Lai, Z.; Liu, C.; Lo, E.; Zhao, Y. Sharding Blockchain. In Proceedings of the 2018 IEEE International Conference on Internet of Things (iThings) and IEEE Green Computing and Communications (GreenCom) and IEEE Cyber, Physical and Social Computing (CPSCom) and IEEE Smart Data (SmartData), Halifax, NS, Canada, 30 July–3 August 2018; p. 1665. [CrossRef]
106. Elrond-A Highly Scalable Public Blockchain via Adaptive State Sharding and Secure Proof of Stake. 2019. Available online: https://elrond.com/assets/files/elrond-whitepaper.pdf (accessed on 24 February 2021).
107. Luu, L.; Narayanan, V.; Zheng, C.; Baweja, K.; Gilbert, S.; Saxena, P. A secure sharding protocol for open blockchains. In Proceedings of the 2016 ACM SIGSAC Conference on Computer and Communications Security, Vienna, Austria, 24–28 October 2016; pp. 17–30. [CrossRef]
108. Kokoris-Kogias, E.; Jovanovic, P.; Gasser, L.; Gailly, N.; Syta, E.; Ford, B. Omniledger: A secure, scale-out, decentralized ledger via sharding. In Proceedings of the 2018 IEEE Symposium on Security and Privacy (SP), San Francisco, CA, USA, 20–24 May 2018; pp. 583–598. [CrossRef]
109. Zamani, M.; Movahedi, M.; Raykova, M. Rapidchain: Scalability blockchain via full sharding. In Proceedings of the 2018 ACM SIGSAC Conference on Computer and Communications Security, Toronto, ON, Canada, 15–19 October 2018; pp. 931–948. [CrossRef]
110. Wilkinso, S.; Boshevski, T.; Brandoff, J.; Prestwich, J.; Hall, G.; Gerbes, P.; Hutchins, P.; Pollard, C. Storj: A Peer-to-Peer Cloud Storage Network. Available online: https://storj.io/storj.pdf (accessed on 24 February 2021).

111. Benet, J. IPFS —Content Addressed, Versioned, P2P File Systems. Available online: https://ipfs.io/ipfs/QmR7GSQM93Cx5eAg6a6yRzNde1FQv7uL6X1o4k7zrJa3LX/ipfs.draft3.pdf (accessed on 24 February 2021).
112. Labs, P. Filecoin: A Decentralized Storage Network. Available online: https://filecoin.io/filecoin.pdf (accessed on 24 February 2021).
113. MediaChain. Available online: http://www.mediachain.io/ (accessed on 24 February 2021).
114. Sevcik, J. DECENT Whitepaper. 2015. Available online: https://www.allcryptowhitepapers.com/decent-whitepaper/ (accessed on 24 February 2021).
115. Vorick, D. Luke Champine, Sia: Simple Decentralizes Storage. Available online: https://sia.tech/sia.pdf (accessed on 24 February 2021).
116. Nick, L.; Ma, Q.; Irvine, D. Safecoin: The Decentralised Network Token. Maidsafe. *Tech. Rep.* **2015**. Available online: https://docs.maidsafe.net/Whitepapers/pdf/Safecoin.pdf (accessed on 24 February 2021).
117. Trón, V.; Fischer, A.; Nagy, D.A.; Felföldi, Z.; Johnson, N. Swap, Swear and Swindle Incentive System for Swarm. Available online: https://ethersphere.github.io/swarm-home/ethersphere/orange-papers/1/sw%5E3.pdf (accessed on 24 February 2021).
118. Arweave. Available online: https://github.com/ArweaveTeam/arweave (accessed on 24 February 2021).
119. Ozyilmaz, K.R.; Yurdakul, A. Designing a Blockchain-Based IoT With Ethereum, Swarm, and LoRa: The Software Solution to Create High Availability With Minimal Security Risks. *IEEE Consum. Electron. Mag.* **2019**, *8*, 28–34. [CrossRef]
120. Wang, S.; Li, G.; Yao, X.; Zeng, Y.; Pang, L.; Zhang, L. A Distributed Storage and Access Approach for Massive Remote Sensing Data in MongoDB. *ISPRS Int. J. Geo-Inf.* **2019**, *8*, 533. [CrossRef]
121. Back, A.; Corallo, M.; Dashjr, L.; Friedenbach, M.; Maxwell, G.; Miller, A.; Poelstra, A.; Timón, J.; Wuille, P. Enabling Blockchain Innovations with Pegged Sidechains. 2014. Available online: https://blockstream.com/sidechains.pdf (accessed on 24 February 2021).
122. Back, A.; Maxwell, G. Transferring Ledger Assets between Blockchains via Pegged Sidechains. U.S. Patent Application No. 15/150,032, 10 November 2016. Available online: https://patents.google.com/patent/US20160330034A1/en (accessed on 24 February 2021).
123. Fallis, A. Rootstock Platform: Bitcoin Powered Smart Contracts—White Paper. *J. Chem. Inf. Model* **2013**, *53*, 1689–1699.
124. Joseph, P.; Dryja, T. The Bitcoin Lightning Network: Scalable off-Chain Instant Payments. 2016. Available online: https://lightning.network/lightning-network-paper.pdf (accessed on 24 February 2021).
125. Raiden. Available online: https://raiden.network/ (accessed on 24 February 2021).
126. Bolt. Available online: https://boltlabs.tech/ (accessed on 24 February 2021).
127. Siris, V.A.; Dimopoulos, D.; Fotiou, N.; Voulgaris, S.; Polyzos, G.C. Decentralized authorization in constrained IoT environments exploiting interledger mechanisms. *Comput. Commun.* **2020**, *152*, 243–251. [CrossRef]
128. Provable. Available online: http://provable.xyz/ (accessed on 24 February 2021).
129. Jason, T.; Reitwießner, C. A Scalable Verification Solution for Blockchains. 2017. Available online: https://people.cs.uchicago.edu/~teutsch/papers/truebit.pdf (accessed on 24 February 2021).
130. WolframAlpha. Available online: https://www.wolframalpha.com/ (accessed on 24 February 2021).
131. Peterson, J. Augur: A Decentralized Oracle and Prediction Market Platform. *arXiv* **2015**, arXiv:1501.01042. Available online: https://arxiv.org/abs/1501.01042 (accessed on 24 February 2021).
132. Haas, A.; Rossberg, A.; Schuff, D.L.; Titzer, B.L.; Holman, M.; Gohman, D.; Bastien, J.F. Bringing the web up to speed with WebAssembly. In Proceedings of the 38th ACM SIGPLAN Conference on Programming Language Design and Implementation, Barcelona, Spain, 18–23 June 2017.
133. Zyskind, G.; Nathan, O. Decentralizing privacy: Using blockchain to protect personal data. In Proceedings of the 2015 IEEE Security and Privacy Workshops, San Jose, CA, USA, 21–22 May 2015; pp. 180–184. [CrossRef]
134. Enigma- Testnet. Available online: https://github.com/enigmampc?language=javascript (accessed on 24 February 2021).
135. AHrga; Capuder, T.; Žarko, I.P. Demystifying Distributed Ledger Technologies: Limits, Challenges, and Potentials in the Energy Sector. *IEEE Access* **2020**, *8*, 126149–126163. [CrossRef]
136. Li, D.; Wong, W.E.; Guo, J. A Survey on Blockchain for Enterprise Using Hyperledger Fabric and Composer. In Proceedings of the 2019 6th International Conference on Dependable Systems and Their Applications (DSA), Harbin, China, 3–6 January 2020; pp. 71–80. [CrossRef]
137. The Interledger Protocol. Available online: https://interledger.org/rfcs/0027-interledger-protocol-4/ (accessed on 24 February 2021).
138. Le, D.; Yang, G.; Ghorbani, A. A New Multisignature Scheme with Public Key Aggregation for Blockchain. In Proceedings of the 17th International Conference on Privacy, Security and Trust (PST), Fredericton, NB, Canada, 26–28 August 2019; pp. 1–7. [CrossRef]
139. Rajan, D.; Visser, M. Quantum Blockchain Using Entanglement in Time. *Quantum Rep.* **2019**, *1*, 2. [CrossRef]
140. Otsuki, K.; Banno, R.; Shudo, K. Quantitatively Analyzing Relay Networks in Bitcoin. In Proceedings of the 2020 IEEE International Conference on Blockchain (Blockchain), Rhodes Island, Greece, 2–6 November 2020; pp. 214–220. [CrossRef]
141. Dai, W.; Deng, J.; Wang, Q.; Cui, C.; Zou, D.; Jin, H. SBLWT: A Secure Blockchain Lightweight Wallet Based on Trustzone. *IEEE Access* **2018**, *6*, 40638–40648. [CrossRef]
142. DriveChain: Enabling Bitcoin Sidechains. Available online: http://www.drivechain.info/ (accessed on 24 February 2021).

 *future internet*

Perspective

The Machine-to-Everything (M2X) Economy: Business Enactments, Collaborations, and e-Governance

Benjamin Leiding [1,*,†], Priyanka Sharma [1,†] and Alexander Norta [2,†]

1. Institute for Software and Systems Engineering, Clausthal University of Technology, 38678 Clausthal-Zellerfeld, Germany; priyanka.sharma@tu-clausthal.de
2. Department of Software Science, Tallinn University of Technology, 12616 Tallinn, Estonia; alexander.norta@taltech.ee
* Correspondence: benjamin.leiding@tu-clausthal.de
† These authors contributed equally to this work.

Abstract: Nowadays, business enactments almost exclusively focus on human-to-human business transactions. However, the ubiquitousness of smart devices enables business enactments among autonomously acting machines, thereby providing the foundation for the machine-driven Machine-to-Everything (M2X) Economy. Human-to-human business is governed by enforceable contracts either in the form of oral, or written agreements. Still, a machine-driven ecosystem requires a digital equivalent that is accessible to all stakeholders. Additionally, an electronic contract platform enables fact-tracking, non-repudiation, auditability and tamper-resistant storage of information in a distributed multi-stakeholder setting. A suitable approach for M2X enactments are electronic smart contracts that allow to govern business transactions using a computerized transaction protocol such as a blockchain. In this position paper, we argue in favor of an open, decentralized and distributed smart contract-based M2X Economy that supports the corresponding multi-stakeholder ecosystem and facilitates M2X value exchange, collaborations, and business enactments. Finally, it allows for a distributed e-governance model that fosters open platforms and interoperability. Thus, serving as a foundation for the ubiquitous M2X Economy and its ecosystem.

Keywords: blockchain; smart contract; M2X; smart autonomous devices; e-governance; lifecycle management

Citation: Leiding, B.; Sharma, P.; Norta, A. The Machine-to-Everything (M2X) Economy: Business Enactments, Collaborations, and e-Governance. *Future Internet* **2021**, *13*, 319. https://doi.org/10.3390/fi13120319

Academic Editor: Paolo Bellavista

Received: 7 December 2021
Accepted: 13 December 2021
Published: 19 December 2021

Publisher's Note: MDPI stays neutral with regard to jurisdictional claims in published maps and institutional affiliations.

Copyright: © 2021 by the authors. Licensee MDPI, Basel, Switzerland. This article is an open access article distributed under the terms and conditions of the Creative Commons Attribution (CC BY) license (https://creativecommons.org/licenses/by/4.0/).

1. Introduction

An open Machine-to-Everything (M2X) Economy [1] emerges when humans and smart autonomous devices interact, transact, and collaborate, e.g., self-driving buses and autonomous food delivery in a smart-city context [2,3]. The ubiquitousness of smart devices also allows for business transactions without human intervention among autonomously acting machines. Besides Machine-to-Machine (M2M) interactions, machines interact with humans (Machine-to-Human–M2H), or infrastructure components (Machine-to-Infrastructure–M2I)—combined they provide the foundation for the machine-driven M2X Economy. While related concepts such as the Internet of Things (IoT), Smart Homes as well as Smart Cities [4], and the Industry 4.0 [5] have evolved, they do not support an interoperable, integrated, scalable model that facilitates the M2X Economy. Likewise, concepts for M2X value transfer, collaborations, and distributed e-governance are missing to achieve the shared objectives. Moreover, integrating humans and smart devices into a well-functioning socio-technical system [6] is essential, as it puts the M2X concept in a human-centered context.

In the M2X Economy, smart sensors may offer collected sensor data such as temperature, or air contamination to interested buyers that rely on the aforementioned data for their own computations. In the context of autonomous and self-driving vehicles, scenarios such as automated tollbooth payments, autonomous battery charging services as well as general

Transportation-as-a-Service (TaaS) applications are among the most discussed use cases [7]. Thus, a socio-technical business model is required as it facilitates the M2X Economy.

Various M2X-resembling applications and use cases already exist, e.g., in the context of IoT. However, complex and impactful applications are still missing that provide more than marginal value to society. In addition, an economy emerging from M2X enactments among humans, smart devices, software agents, and physical systems is rarely considered. To provide or utilize non-trivial services, smart devices may also have to collaborate on-demand with other entities to be able to achieve a shared goal, or even migrate to different geographical locations based on supply and demand. Accordingly, "the interleaved on-demand collaborations, interactions, and transactions among autonomous, heterogeneous, and highly dynamic entities (humans, machines, software agents, etc.) lead to a decentralized, distributed and heterogeneous socio-technical system consisting of a large number of micro-services of different vendors and solution as well as infrastructure providers" [1].

This trend coincides with the emergence of smart-contract blockchain technology [8] that allows for novel peer-to-peer (P2P) electronic governance models. Traditionally, human-to-human business enactments are governed by contracts either in the form of oral, or written agreement. A machine-driven ecosystem requires a digital equivalent that is accessible to all stakeholders, i.e., a smart contract-driven platform that allows for fact tracking, non-repudiation, auditability, and tamper-resistant storage of information in a distributed multi-stakeholder setting. Electronic smart contracts enable and govern business transactions using a computerized transaction protocol such as a blockchain. Moreover, smart-contract blockchain technology comprises computer programs for the consistent execution by a network of mutually distrusting nodes where no arbitration of a trusted authority exists.

A one-stop platform for the provision and enactment of services and goods of a M2X ecosystem is desirable instead of a manufacturer-focused platform with deliberately forced, or functional lock-ins that lead to the formation of self-contained data and service silos such as Tesla, Google, or Amazon. Instead, an interoperabilty layer that implements the compatibility of different manufacturer platforms is required to allow for the exploitation of economies of scale and increased efficiency. Thus providing the foundation for an ecosystem that can be operated as a joint venture of various stakeholders and includes built-in e-governance mechanisms, thereby constituting a neutral territory for all stakeholders.

In this position paper, we argue in favor of an open, decentralized and distributed smart-contract-based M2X Economy that supports the corresponding multi-stakeholder ecosystem and facilitates M2X value exchange, collaborations, and business enactments. Furthermore, the M2X Economy allows for a distributed e-governance model that fosters open platforms and interoperabilty. To do so, we draw from a variety of previous work and assemble an initial set of essential building blocks for a future M2X Economy and its corresponding ecosystem.

The research methodology of this work follows the usual approach of a position paper: First, we stipulate our position by presenting an innovative hypotheses—as stated above, we argue in favor of an open, decentralized, and distributed smart-contract-based M2X Economy. Subsequently, related background information pertaining to the position are provided. Second, we provide evidence to support our position. Third, we follow a discussion of both sides of the matter before concluding the presented position statement.

Our position paper provides three main contributions: First, it is a call for a discussion of an emerging machine-driven economy and its corresponding ecosystem with autonomously acting devices offering and consuming services in a M2X context. Second, it suggests a course of actions for developing the M2X Economy needs to focus on specific domains. Third, it outlines enabling concepts of the M2X Economy.

The remainder of this paper is structured as follows: Section 2 introduces the M2X Economy in detail, showcases the state of the art, and discusses related work. Next, Section 3 focuses on mechanisms for M2X stakeholders to interact, transact, and collaborate by means of a smart-contract-based lifecycle approach and a corresponding distributed e-governance infrastructure. Section 4 details the smart token economics. Subsequently, Section 5 discusses our position as well as alternative approaches. Finally, Section 6 concludes our work.

2. The M2X Economy

The evolving M2X applications and the corresponding ecosystem will influence our daily lives in many ways. Besides M2M interactions, machines interact with humans (M2H) or infrastructure components (M2I). The framework of the M2X Economy represents a more general view on use cases that involve autonomous smart devices and also encompasses M2M, M2H, and M2I scenarios [1].

In Section 2.1 we first present the running case that is used for illustration purposes throughout this work. Afterwards, Section 2.2 introduces related concepts such as cybernetics, IoT, cyber-physical systems (CPS), and wireless sensor networks (WSNs) as well as related work. Next, is the definition and elements of the M2X Economy in Section 2.3.

2.1. Running Case

We introduce an example running case of the M2X Economy in order to provide the reader with a better understanding as well as the scope of M2X applications. The selected running case is illustrated in Figure 1 and belongs to the sub-set of vehicle-focused M2X applications, i.e., the vehicle-to-everything (V2X).

In the future, people might not possess vehicles any more. Instead, vehicles may own themselves, or they are owned by the government, or private corporations [1]. We assume that Alice requests a self-driving car (TaaS) to go from Point *A* to *B* and several route options exist for this. Figure 1 indicates that the fastest route option is expensive but also the most comfortable and equipped with toll gates. Alternatively, the less comfortable, cheaper option is via Point *C* and includes traffic lights and traffic congestion. Alice may select her preferred option depending on her price range and on the urgency of reaching Point *B*. Furthermore, we assume that the self-driving cars are able to communicate with each other as well as the traffic lights (infrastructure). It is also possible to buy a green-light phase for a faster commute to Point *B*. Finally, Figure 1 shows an electric charging station near Point *B* that the self-driving cars may use for some amount of fee. In the described running case, assuming that time and money are important factors, Alice may select from a range of possible options. On the one hand, she may choose the fastest and most expensive route to Point *B*, or take the less comfortable and cheaper option via Point *C*. Additionally, she can pay an extra fee and her car may negotiate for a green light at the traffic signals.

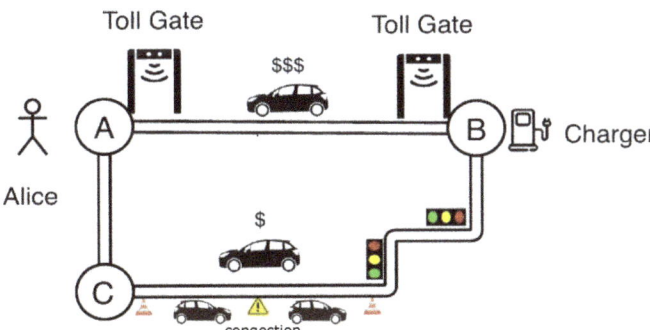

Figure 1. Self-driving M2X running case incorporating smart traffic lights and a traffic-congestion response, adapted from [1].

Our running case—despite it simplicity—already covers a wide variety of M2X service enactments, i.e., TaaS, toll gate payments, battery electric vehicle (BEV) charging, road space negotiations, smart parking, and traffic information provision. Nevertheless, they also only constitute a small subset of services within the M2X ecosystem.

2.2. State of the Art and Related Work

The idea of the M2X Economy and its ecosystem overlaps with some closely related concepts and applications such as cybernetics, WSNs, CPS, and IoT [1]. This section clarifies the differences and overlaps with those concepts and applications.

Wiener [9] defines the concept of cybernetics as "the scientific study of control and communication in the animal and the machine", while WSNs consist of spatially distributed autonomous sensors to monitor physical or environmental conditions and to cooperatively pass their data through a variety of networks to a main location [10].

CPS are engineered systems that are built from, and depend upon, the seamless integration of computation and physical components. CPS tightly integrate computing devices, actuation, and control, networking infrastructure, and sensing of the physical world [11].

Gubbi et al. [12] defines IoT as an "interconnection of sensing and actuating devices providing the ability to share information across platforms through a unified framework, developing a common operating picture for enabling innovative applications. This is achieved by seamless, large-scale sensing, data analytic and information representation using cutting edge ubiquitous sensing and cloud computing".

Robotic Process Automation (RPA) is regarded as one of the most advanced technologies in the area of computers science, electronic and communications, mechanical engineering, and information technology [13]. With software robots autonomously executing their choreography uninterruptedly, quickly, and flawlessly while at the same time being easy to implement at relatively low costs compared to traditional process automation, RPA may automate processes enabling business transactions in the near future [14].

After clarifying the terms and concepts above, the question remains: Where does the M2X Economy fit in? Several publications list and survey CPS and IoT applications, e.g., [15–19]), as well as their economic value and impact, e.g., [19–21]. However, the emerging economy resulting from M2X enactments among humans, smart devices, software agents and physical systems is rarely considered.

2.3. Elements and Definition of the M2X Economy

The M2X Economy framework involves autonomous smart devices and further encompasses mobile devices, software agents, humans, and infrastructure in M2M, M2H, and M2I scenarios. A main requirement of such an ecosystem is to enable a seamless integration of humans and smart devices into a well functioning socio-technical system that puts the M2X concept in a human-centered context [1]. When considering collaborations and interactions between the M2X stakeholders, multilevel and unidirectional interrelations can be seen. The interleaved on-demand collaborations, interactions and transactions among autonomous, heterogeneous and highly dynamic entities (humans, machines, software agents, etc.) lead to decentralized and distributed socio-technical systems comprising a large number of micro-services of different vendors and solutions, as well as infrastructure providers [1].

Definition 1. *Thus, the M2X Economy is the result of interactions, transactions, collaborations and business enactments among humans, autonomous and cooperative smart devices, software agents, and physical systems. The corresponding ecosystem is formed by automated, globally-available, heterogeneous socio-technical e-governance systems with loosely coupled, P2P-resembling network structures and is characterized by its dynamic, continuously changing, interoperable, open and distributed nature. Thereby, the M2X Economy employs concepts such as cyber-physical systems, the Internet of Things, and wireless sensor networks.*

3. Enactment, Collaboration, and e-Governance

Human-to-human business enactments are governed by enforceable contracts either in the form of an oral, or written agreement. Contract documents [22] uniquely identify the contracting parties, the offered services, or goods, a corresponding compensation, as well as further constraints such as delivery dates, quality goals, penalties, and means of arbitration [23]. Still, a highly-automated and machine-driven ecosystem requires a digital equivalent that is accessible to and usable by all stakeholders. Moreover, traditional solely human-focused contracts are often under-specified and thus, not suitable for M2X enactments [23]. "Most importantly, traditional contracts do not provide sufficient details about the actual transaction process, and consequently, frictions between the contracting parties are very likely, e.g., one party assumes a specific product certificate before delivering a partial compensation, and the other party assumes the opposite" [23].

Electronic smart contracts [24,25] address the listed issues by enabling and governing business transactions using a computerized transaction protocol such as a blockchain. Blockchain technology [26] ensures a trustworthy, tamper-resistant, P2P transaction processing, and enables a distributed, often decentralized, transparent way for communication. More generally, a blockchain is a distributed ledger that enables users to send data, process it, and verify it without the need for a central entity [26]. In addition, smart-contract blockchain technology comprises computer programs for the consistent execution by a network of mutually distrusting nodes where no arbitration of a trusted authority exists. As a result, allowing for fact tracking, non-repudiation, auditability, and tamper-resistant storage of information in a distributed multi-stakeholder setting.

On the one hand, the running case of Section 2.1 only presents a small fraction of potential applications and use cases of the M2X Economy. On the other hand, the running case already contains several examples of different M2X interactions, transactions, and collaborations, i.e., TaaS, road space negotiations, toll gate payments, BEV charging, traffic light information dissemination, and smart parking. The enactments of the listed examples follow a similar process structure, thus allowing for an abstraction towards a general lifecycle of the M2X Economy. Consequently, we stipulate that all M2X-related interactions, transactions, collaborations, and further enactments can be governed and represented using a blockchain-based smart contract.

In the following, Section 3.1 details a conceptual lifecycle for M2X business enactments and collaborations using electronic smart contracts. Afterward, Section 3.2 outlines corresponding distributed e-governance mechanisms.

3.1. Digital Contract Lifecycle Management

Based on [23], Norta presents a conceptual smart contract-based lifecycle as illustrated in Figure 2.

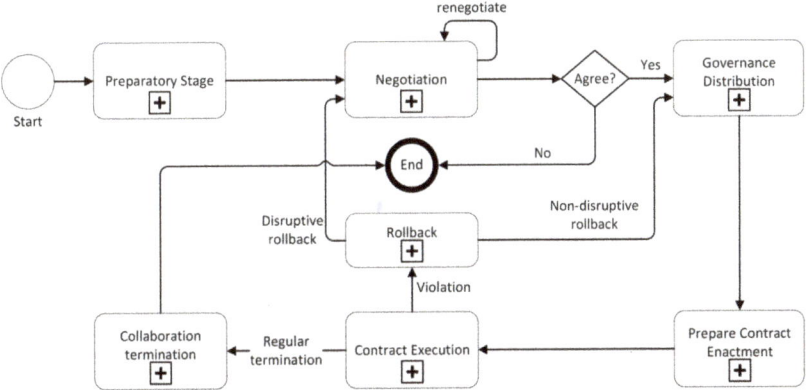

Figure 2. Conceptual lifecycle for M2X business enactments–Based on [1,23].

The lifecycle is divided into seven stages: (*i.*) preparation, (*ii.*) negotiation, (*iii.*) governance distribution (*iv.*) preparation of collaboration enactment (*v.*) collaboration enactment (*vi.*) rollback, and (*vii.*) termination stage.

The preparatory stage is initiated by selecting a pre-configured template from a distributed service hub. The distributed service hub hosts contract templates that match different M2X use-cases and outlines the corresponding contractual process flow. Following the running case, a template for TaaS is selected and populated with information about the involved entities, such as identifiers and wallet addresses. Moreover, TaaS-specific conditions are defined, e.g., departure location, final destination, the required vehicle size, and the departure/arrival time. Subsequently, the TaaS contract request is negotiated with potential TaaS service providers, i.e., autonomous vehicles. The negotiated-contract conditions primarily depend on information such as the travel distance and energy consumption of the vehicle as well as the number of transported individuals.

The negotiation stage concludes either with an agreement—resulting in a contract signed by both parties to express their approval—or a contract rollback if no agreement is reached. In our case, Alice and the vehicle serving the direct route between A and B agree upon a set of rights and obligations. Subsequently, a smart contract is established and serves as a distributed governance infrastructure (DGI) coordinating agent (also see Figure 3). Finally, the e-governance distribution commences, Alice and the vehicle each receive local contract copies containing the respective obligations and rights of each party resulting from the previous negotiations [23]. The vehicle's and Alice's obligations are observed by monitors and are assigned so-called business-network model agents (BNMA) that connect to IoT-sensors such as the vehicle's GPS-sensor [23].

The required process endpoints, e.g., for payment processing as Alice pays using the cryptocurrency of her choice, are prepared and provided as part of the contract enactment preparation. "Once the e-governance infrastructure is set up, technically realizing the behavior in the local copies of the contracts requires concrete local electronic services. After picking these services, follows the creation of communication endpoints so that the services of the partners are able to communicate with each other. The final step of the preparation is a liveness check of the channel-connected services" [23].

Next, the contract execution stage is triggered, and the vehicle picks up Alice at location A. The TaaS contract enactment terminates, or expires once Alice arrives at Point B. Alternatively, the contract is prematurely terminated, e.g., failing to transport Alice to Point B, or violating agreed upon time restrictions, might result in an immediate rollback of the TaaS contract, or invokes a mediation process that is supervised by a conflict-resolution escrow service that is not depicted in Figure 2. Note that the enactment of the TaaS running case subsumes further M2X enactments that occur throughout the TaaS service provision, e.g., the vehicle pays a minor fee at the toll gate to use the faster toll road. The toll road payment is part of the costs to transport Alice from Point A to B and is thus, included in her fare.

3.2. Distributed e-Governance

While Figure 2 presents the collaboration among partners from a lifecycle perspective, Figure 3 depicts the creation sequence of a DGI from an infrastructure perspective, thereby providing the foundation for a distributed, interoperable, dynamic ad-hoc enactment among heterogeneous M2X entities.

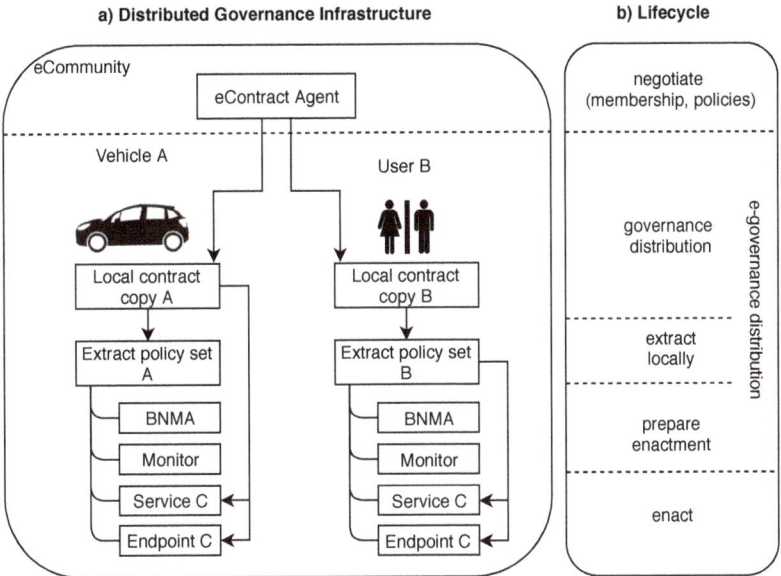

Figure 3. Distributed M2X governance infrastructure. Source: [1] and based on [23,27].

Finally, the M2X collaboration model enables providers to decide if and in which way changes to a private and internal process must be projected to a related public process view in a way where the process view and the internal process stay consistent with each other. Thus, the M2X collaboration model enables service-consumers to monitor a public process view to safely follow changes performed to a private and internal process.

This way, it is possible to support the evolution of smart contracts [28] as a significant means to achieve flexibility in B2B collaborations. As smart contracts are instrumental to enable decentralized autonomous organizations (DAO) [23] for the formation of electronic communities, service-oriented cloud computing (SOCC) [29] supports companies in the coordination of information- and business-process flows [30] for the choreography and orchestration [31] of heterogeneous legacy-system infrastructures.

For evolving DAO-collaborations, Figure 4a shows a conceptually collaboration configuration where the template for an electronic-community formation is given by a business-network model (BNM) [32] to specify choreographies relevant for a respective business scenario. The BNM defines legally valid [33–35] template contracts as service types together with assigned organizational roles. A collaboration hub that houses business processes as a service (BPaaS-HUB) [36] in the form of process views [30], houses the BNM templates for potential collaborating counterparties to enable a speedy matching.

The external layer of Figure 4a depicts service offers to identically match the service types defined in the BNM with the respective collaborating partner contractual sphere. Furthermore, a collaborating partner is required to comply with a specific partner roles assigned to a specific service type. In [30], further details are contained about a tree-based process-view matching for creating DAO-configurations. We stress that Figure 4a uses Petri net [37] notation, which can be mapped into a tree-formalization as well with less computationally expensive strain.

Figure 4b presents a corresponding mapping and presents the top-level structure of a smart contract using the eSourcing Markup Language (eSML) [38]. "The core structure of a smart contract we organize according to the interrogatives *Who* for defining the contracting parties together with their resources and data definitions, *Where* to specify the business and legal context, and *What* for specifying the exchanged business values. For achieving a consensus, we assume the What-interrogative employs matching process views that require cross-organizational alignment for monitorability" [23].

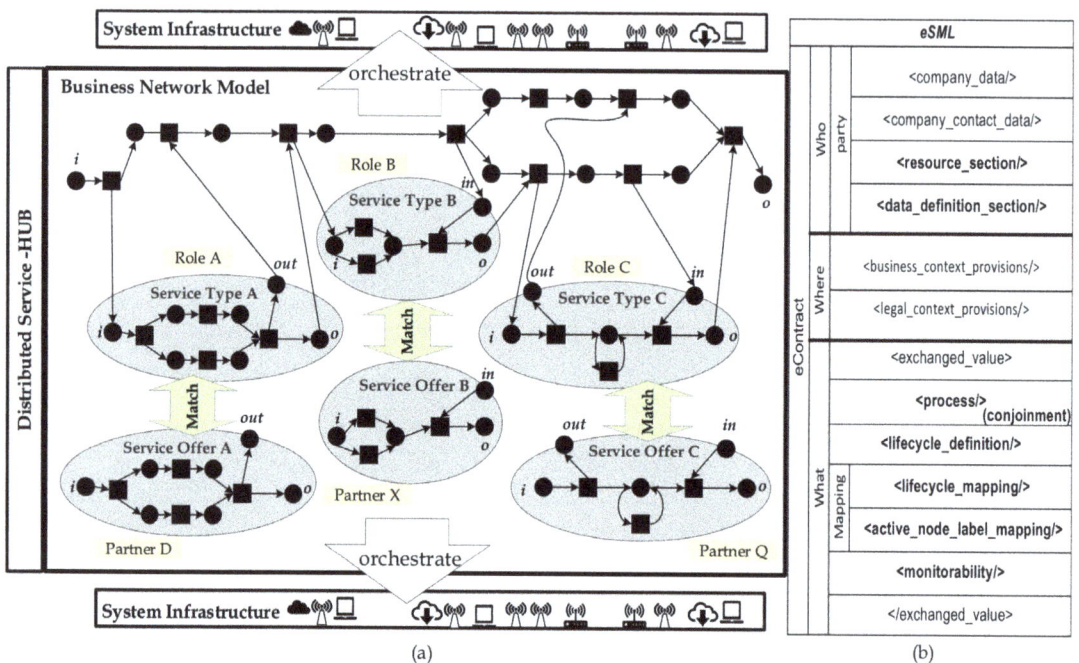

Figure 4. (a) P2P service matching and provision of the M2X ecosystem using the eSourcing framework–(Based on) [23]. (b) The eSourcing Markup Language (eSML) for specifying contractual collaborations–Based on [38].

4. Smart Token Economics

The running case of Section 2.1 shows that the M2X Economy is a complex, distributed, and socio-technical framework that requires a novel approach for developing the monetary economy. We infer that the traditional financial system is not suitable and lacks the utility for consideration in the M2X Economy. An important reason is that an integration of the financial legacy technology does not scale and perform for a context such as the running case in Figure 1 and additionally, to technically support the incentives mechanisms between the human user termed Alice and the smart autonomous devices being the cars, traffic lights, toll gates, and charging stations, we require programmable monetary units, which fiat-currencies are not, e.g, as a code extension of an ERC20-token smart-contract template (https://eips.ethereum.org/EIPS/eip-20 accessed on 4 November 2021). Consequently, the novel domain of token economics [39] emerges to compensate for the deficiencies of the legacy fiat-currency system. Informally, a token economy in an M2X Economy that employs smart-contract blockchain technology, is characterised by encouraging desirable behavior by the human and artificial agents and infrastructure involved by offering rewards and optionally also penalties in the form of crypto tokens.

We stress that established schools of thought of economics do not typically assume that a monetary unit is programmable and connected as such to a socio-technical application system context as Section 2.1 describes, where the automated complex governance of incentives mechanisms is essential for P2P interactions between humans, smart autonomous devices, and infrastructure. On the other hand, a set of standard-token smart contracts are available, initially offered by Ethereum, that allow for flexible instantiations into diverse token types [40], e.g., tokens for a platform, that play a role of a security, or facilitate transactions, enable specific platform-utility use, e-governance tokens for complex voting mechanisms, reputation tokens, and so on (https://tinyurl.com/token-types accessed on 10 November 2021).

As token economics based on smart-contract blockchain technology is an emerging computer-science driven scientific discipline, we infer that the programmable nature of crypto tokens requires a novel development methodology that is integrated with the M2X system design from the very inception. In earlier research [41], we discover that no suitable methodology exists for developing blockchain distributed applications (DApps), which is relevant too for an M2X context. Consequently, the distributed agent-oriented modeling (DAOM) method [42] fills this gap, being the first blockchain-DApp development method that also integrates the foundation for the development of a DApp-specific token economy being integrated with the system functionalities.

While due to page limitations, we refer interested readers to several use cases [43,44], where the DAOM method follows a set of briefly described model-driven design steps. First, the functional and quality goals, together with human and artificial software agents are organized into a so-called goal model where transparent gray rectangles with token-type labels denote smart-contract blockchain application in a DApp. Next, based on a set of heuristics, a component-diagram architecture is deduced from the goal model where blockchain-involving components are also gray colored, corresponding to the specific requirements of derivation. The addition in the component-diagram architecture is the specification of the information-exchange channels between components, and components to human and artificial software agents. Based on this conceptual DApp understanding, DAOM next prescribes the specification of so-called on-chain transaction sets that are a tuple comprising an ID, short description and agents involved per respective transaction evaluation. It is important to specify this on-chain transaction set given the expenses of transaction validations [45], e.g., per proof-of-work (PoW), proof-of-staking (PoS), and so on. Finally, the set of information-exchange protocols between components, and components with human and artificial software agents, is expressed either in sequence diagrams, or in a graph-based notation such as business process model and notation (BPMN) [46] in which the IDs of respective on-chain transactions are embedded.

Note that the DAOM method is inherently technology agnostic and allows subsequently for deducing a technology stack with a considerable blockchain subset for a detailed token-economics establishment to govern the incentive mechanisms and a rapid Dapp development. At the same time, extension work is required to develop DAOM further for full applicability in an M2X context. More concretely, since smart autonomous devices are an essential part of M2X, being software agents embedded in hardware, further modeling notations must be adopted into the DAOM method for designing specifically the behavior of the P2P-communicating smart autonomous devices and also the smart-contract instantiations that constitute the respective token types to govern the incentive mechanisms. A promising option is to consider agent-based computational economics [47] in combination with a future extended DAOM method for M2X-focused smart-token economics development.

5. Discussion

The previous Section 2 introduces the M2X Economy, while Sections 3 and 4 focus on essential building blocks of the M2X Economy, i.e., M2X enactments, governance and smart-token economics. Subsequent sections discuss the arguments in favor and against our smart-contract enabled and blockchain-based M2X proposal as well as alternative approaches. Space constraints force us to focus on the most relevant aspects.

5.1. Digital Smart Contracts

While human-to-human business enactments are governed by oral, or written contracts, they are not applicable to the highly-automated, machine-driven and human-focused M2X Economy. First, human-centered oral and written contracts are difficult to process even for smart machines [1]. Second, traditional contracts [48] are often under-specified and do not provide sufficient details about the actual transaction processes as well as about the parties obligations and rights [23,34]. Third, they do not allow for extensive automation, scale badly and lack a computerized transaction protocol [49]. Fourth, efficient and automated means of conflict-resolution are missing [1,23].

While we propose the utilization of electronic smart contracts to address the issues above, one may argue that a cloud-based online shop for services of the M2X Economy would be sufficient, e.g., Amazon's web shop proves to scale well and even partially automates business enactments. Still, such types of business enactments suffer from transparency issues which complicate—or even prevent and sabotage—conflict-resolution mechanisms. Especially the unequal power relations between a single entity and the service-offering cloud shop prevent fair markets and business enactments.

In contrast, smart contracts allow for the automated, consistent, transparent, and auditable enactment of contracts by a network of mutually distrusting nodes where no arbitration of a trusted authority is required [24,50,51]. As a result, allowing for fact tracking, non-repudiation, auditability, and tamper-resistant storage of information in a distributed multi-stakeholder setting. In case of any conflicts, pre-defined rollback mechanisms are applied as described in [23].

Finally, Amazon-resembling service provision promotes lock-in effects, and obstructs much needed interoperability and openness of the M2X ecosystem as discussed in the subsequent Section 5.2. Neither traditional contracts, nor a cloud-hosted shop-resembling service provisions, allow for dynamic, P2P- (even local) ad-hoc enactments.

5.2. Openness and Interoperability

A one-stop platform for the provision and enactment of services and goods of a M2X ecosystem is desirable instead of a manufacturer-focused platform with deliberately forced, or functional lock-ins that lead to the formation of self-contained data and service silos such as Tesla, Google, or Amazon. As suggested in [1], interoperability allows for the exploitation of economies of scale and increased efficiency. At the same time, an interoperable blockchain ecosystem can be operated as a joint venture of various stakeholders and include built-in e-governance mechanisms, thereby constituting a neutral territory for all stakeholders [1,52]. A smart-contract driven M2X platform and its corresponding ecosystem not only enable an interoperable platform for M2X entities, but also further reduces dependency on intermediaries [53].

The technical implementation is realized by so-called relay chains as introduced by Polkadot [52] that provide communication interfaces for different heterogeneous blockchain platforms to interact with each other and subsequently, allow for a blockchain-agnostic, highly-automated, globally-available orchestration and choreography of heterogeneous socio-technical systems. Thus, specific manufacturers, or service-provider specific functionalities may also be accessible outside their own platform.

5.3. Identity

In order for hardware devices, humans and software agents to conduct digital business transactions, or enact digital collaborations as described in Section 2.1, all these entities require a digital representation of their "real-world" identity. To enable secure business collaborations and transaction within the M2X Economy, this digital representation is required to establish and enable trust, reputation mechanisms, perform verifiable and accountable transactions, and establish reliable as well as auditable data provenance [1]. As M2X is a multi-stakeholder ecosystem, the identity management issue applies not only for its users, but also infrastructure providers, OEMs, regulators and service providers. A single central authority for identity management of all these different stakeholders poses the risk of single point of failure. Furthermore, identity silos create privacy concerns and are not interoperable [54].

As earlier argued in this section, centralized infrastructures are not suitable for facilitating the full potential of the M2X ecosystem. Hence, a centralized identity solution is not an option and a decentralized interoperable identity solution is required. In order to prevent the aforementioned flaws and enable an open interoperable ecosystem, the identity-management solution needs to be self-sovereign and user-centric. Self-sovereign identity puts end-users in charge of decisions about their own privacy and disclosure of their personal information and credentials [54] and not the organizations that traditionally centralize identity. Self-sovereign identity systems that are based on decentralized identifiers (DIDs) [55], utilize distributed ledgers, or blockchains as a distributed storage system that replace centralized and incompatible data silos with a cooperative shared storage resource. The result is a user-controlled identity provision model where users control access and sharing of their data based on a need-to-know-basis using the concepts of DIDs, DID documents, and verifiable claims [1].

5.4. Trust

Blockchains are trust engines in an inherently trustless M2X Economy collaboration context. Blockchain technology promises to secure the M2X ecosystem where the management of large and distributed datasets in a secure way is essential. Still, the expected performance and scalability of existing blockchains is currently not compatible for a M2X context [56]. Consequently, new types of blockchains with novel consensus and validation algorithms are required for the large number of securely connected smart autonomous devices that interact with other machines, humans, and infrastructure.

Since M2X ecosystems are a source of large, unstructured data sets that must be combined and understood to extract intelligence with advanced analytic for actionable decision-making, it is our contention that trust management is only possible with novel blockchain technology of high scalability and performance. For example, the use of blockchains in a M2X ecosystem involves many devices that have low storage capacity and computing power. Since these devices cannot maintain a blockchain of many gigabytes, novel sharding management for blockchain parts to and from devices is required to overcome storage and computing-power limitations [1,57].

5.5. Tokenized Value Exchange

A blockchain-based solution enables the decentralized settlement of value added in the form of crypto tokens [26,58]. The latter may be created entirely without trusted third parties, or intermediaries and exchanged directly P2P [53] while at the same time increasing transaction speed. Since Section 4 stipulates that the legacy financial technologies with a focus on fiat currencies is not suitable and lacks the required utility for the M2X Economy, we put forward further arguments that justify the need for a smart-contract blockchain based token economy. Given the legal and socio-technical complexity of a M2X Economy, it is essential to have a flexible monetary instrument that allows for flexibility with respect to defining for a token the application goals, the properties, the business, and incentivizing governance models. Important for the development of a token model with a specific degree of M2X required complexity is to also target in that process the desired legal-compliance adjustment. Certainly for tokens with a high degree of contextual application complexity, e.g., to tackle governance issues in a M2X Economy, the business-model engineering gains in dominance additionally to legal-compliance assurance.

To expand on the topic of e-governance by tokens, essential for this is the provision of a rich and real-time availability of large data sets stemming from the entities that comprise a M2X Economy. Smart-contract blockchain tokens pose via their incentivized transaction involvement that they facilitate the generation of such data with all economic action involved. With all that, the scope emerges for establishing a novel scientific discipline that may be termed economic systems engineering. Thus, diverse economics and engineering disciplines need to be combined in this novel scientific discipline for M2X Economics in which blockchain-specific consensus mechanisms such as PoW allow for a real-time steering of complex governance scenarios in a trustless collaboration context of complex and adaptive M2X Economies where all services are tokenized themselves.

6. Conclusions and Future Work

This position paper argues for a novel business model for the emerging M2X Economy of multi-stakeholders that is open, decentralized, and distributed. As such, the M2X Economy encompasses the interactions between smart autonomous devices with other machines, humans and infrastructure in a cybernetic context. As an example, we correspondingly present a running case from the domain of self-driving autonomous smart vehicles to be rented by humans for transportation on roads with smart toll gates and smart traffic lights in interaction with other smart vehicles.

Important supporting concepts for the M2X Economy are lifecycle management for the setup, establishment, rollout, rollback and orderly termination of business collaborations. This lifecycle manages cross-organizational process-aware collaboration establishment that is expressed in machine-readable smart contracts.

The suggested course of actions for developing the M2X Economy needs to focus on specific domains. First, since smart contracts are a promising means for managing ad-hoc P2P contractual collaboration establishment, it is important to develop smart-contract languages that have legal relevance with their representation in a machine-readable format. Important is in this context that openness and interoperability must be assured to avoid self-contained data silos and instead enable collaboration transparency for effortless conflict-resolution e-governance mechanisms. Next, an M2X Economy requires the adoption of novel identity authentication for the participating entities and humans that are flexible in the adoption of application-context adjusted challenge sets. Thereby considering scalable and highly performing blockchain technology, a trusted entry into and exit from an M2X ecosystem can be assured for smart autonomous devices, machines, infrastructure and humans. Finally, an M2X Economy should have its incentive mechanisms governed by programmable, smart token sets that are developed with means of smart-contract blockchain technologies.

Exploring the solution options, we observe that smart contracts still lack legal relevance due to missing language contracts. For example, traditional contracts are based on the formulation of obligations and rights that should be part of smart contracts in a machine-readable form. To achieve openness and interoperability for an M2X Economy, the lack of standards that technology providers adhere to should be addressed. For addressing the topic of suitable identity-authentication mechanisms, we claim that the investigation of application-context dependent multi-factor challenge sets are a promising means for trusted entries and exits of humans and non-human actors into a M2X ecosystem. A novel generation of blockchains with scaling and performing consensus algorithms is essential to assure effective trust assurance by investigating novel distributed blockchain-sharding management. Finally, the need arises for establishing economic systems engineering as a scientific discipline for investigating the important domain of tokenized M2X value exchanges.

Author Contributions: Conceptualization, B.L., P.S. and A.N.; Writing—original draft, B.L., P.S. and A.N.; Writing—review & editing, B.L., P.S. and A.N. All authors have read and agreed to the published version of the manuscript.

Funding: We acknowledge financial support by Open Access Publishing Fund of Clausthal University of Technology.

Data Availability Statement: Not Applicable, the study does not report any data.

Conflicts of Interest: The authors declare no conflict of interest.

References

1. Leiding, B. The M2X Economy–Concepts for Business Interactions, Transactions and Collaborations Among Autonomous Smart Devices. Ph.D. Thesis, University of Göttingen, Göttingen, Germany, 2020. Available online: http://hdl.handle.net/21.11130/00-1735-0000-0005-12E4-5 (accessed on 30 October 2021).
2. Männi, M. Tallinn's New Self-Driving Bus Emerged from a University Robotics Course. 2020. Available online: https://estonianworld.com/technology/tallinns-new-self-driving-bus-emerged-from-a-university-robotics-course (accessed on 29 September 2021).
3. Bellan, R. Starship Technologies Is Bringing Food Delivery Robots to Four More US College Campuses This Year. 2021. Available online: https://techcrunch.com/2021/08/10/starship-technologies-is-bringing-food-delivery-robots-to-four-more-us-college-campuses-this-year (accessed on 18 October 2021).
4. Lynggaard, P.; Skouby, K. Complex IoT Systems as Enablers for Smart Homes in a Smart City Vision. *Sensors* **2016**, *16*, 1840. [CrossRef] [PubMed]
5. Vaidya, S.; Ambad, P.; Bhosle, S. Industry 4.0–A Glimpse. *Procedia Manuf.* **2018**, *20*, 233–238. [CrossRef]
6. Savaget, P.; Geissdoerfer, M.; Kharrazi, A.; Evans, S. The Theoretical Foundations of Sociotechnical Systems Change for Sustainability: A Systematic Literature Review. *J. Clean. Prod.* **2019**, *206*, 878–892. [CrossRef]
7. Leiding, B.; Vorobev, W.V. Enabling the V2X Economy Revolution Using a Blockchain-based Value Transaction Layer for Vehicular Ad-hoc Networks. In Proceedings of the 12th Mediterranean Conference on Information Systems–MCIS, Corfu, Greece, 28–30 September 2018.
8. Udokwu, C.; Kormiltsyn, A.; Thangalimodzi, K.; Norta, A. The State of the Art for Blockchain-Enabled Smart-Contract Applications in the Organization. In Proceedings of the 2018 Ivannikov Ispras Open Conference (ISPRAS), Moscow, Russia, 22–23 November 2018; pp. 137–144.
9. Wiener, N. *Cybernetics or Control and Communication in the Animal and the Machine*; MIT Press: Cambridge, MA, USA, 2019.
10. Wan, J.; Chen, M.; Xia, F.; Di, L.; Zhou, K. From Machine-to-Machine Communications Towards Cyber-Physical Systems. *Comput. Sci. Inf. Syst.* **2013**, *10*, 1105–1128. [CrossRef]
11. Cyber-Physical Systems (CPS) | NSF-National Science Foundation, nsf19553. 2019. Available online: https://www.nsf.gov/publications/pub_summ.jsp?ods_key=nsf19553 (accessed on 7 September 2021).
12. Gubbi, J.; Buyya, R.; Marusic, S.; Palaniswami, M. Internet of Things (IoT): A Vision, Architectural Elements, and Future Directions. *Future Gener. Comput. Syst.* **2013**, *29*, 1645–1660. [CrossRef]
13. Madakam, S.; Holmukhe, R.M.; Jaiswal, D.K. The Future Digital Work Force: Robotic Process Automation (RPA). *JISTEM-J. Inf. Syst. Technol. Manag.* **2019**, *16*. [CrossRef]
14. Hofmann, P.; Samp, C.; Urbach, N. Robotic Process Automation. *Electron. Mark.* **2020**, *30*, 99–106. [CrossRef]
15. Khaitan, S.K.; McCalley, J.D. Design Techniques and Applications of Cyberphysical Systems: A Survey. *IEEE Syst. J.* **2015**, *9*, 350–365. [CrossRef]
16. Gunes, V.; Peter, S.; Givargis, T.; Vahid, F. A Survey on Concepts, Applications, and Challenges in Cyber-Physical Systems. *KSII Trans. Internet Inf. Syst.* **2014**, *8*, 4242–4268.
17. Shi, J.; Wan, J.; Yan, H.; Suo, H. A Survey of Cyber-Physical Systems. In Proceedings of the 2011 International Conference on Wireless Communications and Signal Processing (WCSP), Nanjing, China, 9–11 November 2011; pp. 1–6.
18. Da Xu, L.; He, W.; Li, S. Internet of Things in Industries: A Survey. *IEEE Trans. Ind. Informatics* **2014**, *10*, 2233–2243.
19. Al-Fuqaha, A.; Guizani, M.; Mohammadi, M.; Aledhari, M.; Ayyash, M. Internet of Things: A Survey on Enabling Technologies, Protocols, and Applications. *IEEE Commun. Surv. Tutorials* **2015**, *17*, 2347–2376. [CrossRef]
20. Monostori, L.; Kádár, B.; Bauernhansl, T.; Kondoh, S.; Kumara, S.; Reinhart, G.; Sauer, O.; Schuh, G.; Sihn, W.; Ueda, K. Cyber-Physical Systems in Manufacturing. *Cirp Ann.* **2016**, *65*, 621–641. [CrossRef]
21. Ivančić, L.; Vugec, D.S.; Vukšić, V.B. Robotic Process Automation: Systematic Literature Review. In Proceedings of the International Conference on Business Process Management, Vienna, Austria, 1–6 September 2019; Springer: Berlin/Heidelberg, Germany, 2019; pp. 280–295.
22. Roxenhall, T.; Ghauri, P. Use of the Written Contract in Long-lasting Business Relationships. *Ind. Mark. Manag.* **2004**, *33*, 261–268. [CrossRef]
23. Norta, A. Designing a Smart-Contract Application Layer for Transacting Decentralized Autonomous Organizations. In Proceedings of the International Conference on Advances in Computing and Data Sciences, Ghaziabad, India, 11–12 November 2016; Springer: Berlin/Heidelberg, Germany, 2016; pp. 595–604.
24. Szabo, N. Smart Contracts. 1994. Available online: http://www.fon.hum.uva.nl/rob/Courses/InformationInSpeech/CDROM/Literature/LOTwinterschool2006/szabo.best.vwh.net/smart.contracts.html (accessed on 2 October 2021).
25. Szabo, N. The Idea of Smart Contracts. 1997. Available online: http://www.fon.hum.uva.nl/rob/Courses/InformationInSpeech/CDROM/Literature/LOTwinterschool2006/szabo.best.vwh.net/smart_contracts_idea.html (accessed on 2 October 2021).
26. Nakamoto, S. Bitcoin: A Peer-to-Peer Electronic Cash System. 2008. Available online: https://bitcoin.org/bitcoin.pdf (accessed on 9 September 2021).
27. Kutvonen, L.; Norta, A.; Ruohomaa, S. Inter-Enterprise Business Transaction Management in Open Service Ecosystems. In Proceedings of the 2012 IEEE 16th International Enterprise Distributed Object Computing Conference, Beijing, China, 10–14 September 2012; pp. 31–40.

28. Eshuis, R.; Norta, A.; Roulaux, R. Evolving process views. *Inf. Softw. Technol.* **2016**, *80*, 20–35. [CrossRef]
29. Badawy, M.M.; Ali, Z.H.; Ali, H.A. QoS provisioning framework for service-oriented internet of things (IoT). *Clust. Comput.* **2020**, *23*, 575–591. [CrossRef]
30. Eshuis, R.; Norta, A.; Kopp, O.; Pitkänen, E. Service outsourcing with process views. *IEEE Trans. Serv. Comput.* **2013**, *8*, 136–154. [CrossRef]
31. Norta, A.; Grefen, P.; Narendra, N.C. A reference architecture for managing dynamic inter-organizational business processes. *Data Knowl. Eng.* **2014**, *91*, 52–89. [CrossRef]
32. Ruokolainen, T.; Ruohomaa, S.; Kutvonen, L. Solving Service Ecosystem Governance. In Proceedings of the 2011 IEEE 15th International Enterprise Distributed Object Computing Conference Workshops, Helsinki, Finland, 29 August–2 September 2011; pp. 18–25.
33. Dwivedi, V.; Pattanaik, V.; Deval, V.; Dixit, A.; Norta, A.; Draheim, D. Legally Enforceable Smart-Contract Languages: A Systematic Literature Review. *ACM Comput. Surv. (CSUR)* **2021**, *54*, 1–34. [CrossRef]
34. Dwivedi, V.; Norta, A.; Wulf, A.; Leiding, B.; Saxena, S.; Udokwu, C. A Formal Specification Smart-Contract Language For Legally Binding Decentralized Autonomous Organizations. *IEEE Access* **2021**, *9*, 76069–76082. [CrossRef]
35. Dwivedi, V.; Norta, A. A Legal-Relationship Establishment in Smart Contracts: Ontological Semantics for Programming-Language Development. In Proceedings of the International Conference on Advances in Computing and Data Sciences, Nashik, India, 23–24 April 2021; Springer: Berlin/Heidelberg, Germany, 2021; pp. 660–676.
36. Norta, A.; Kutvonen, L. A cloud hub for brokering business processes as a service: A "rendezvous" platform that supports semi-automated background checked partner discovery for cross-enterprise collaboration. In Proceedings of the 2012 Annual SRII Global Conference, San Jose, CA, USA, 24–27 July 2012; pp. 293–302.
37. Petri, C.A.; Reisig, W. Petri net. *Scholarpedia* **2008**, *3*, 6477. [CrossRef]
38. Norta, A.; Ma, L.; Duan, Y.; Rull, A.; Kõlvart, M.; Taveter, K. eContractual Choreography-language Properties Towards Cross-organizational Business Collaboration. *J. Internet Serv. Appl.* **2015**, *6*, 1–23. [CrossRef]
39. Lee, J.Y. A Decentralized Token Economy: How Blockchain and Cryptocurrency can Revolutionize Business. *Bus. Horizons* **2019**, *62*, 773–784. [CrossRef]
40. Di Angelo, M.; Salzer, G. Tokens, Types, and Standards: Identification and Utilization in Ethereum. In Proceedings of the 2020 IEEE International Conference on Decentralized Applications and Infrastructures (DAPPS), Oxford, UK, 3–6 August 2020; pp. 1–10.
41. Udokwu, C.; Anyanka, H.; Norta, A. Evaluation of Approaches for Designing and Developing Decentralized Applications on Blockchain. In Proceedings of the 2020 4th International Conference on Algorithms, Computing and Systems, Rabat, Morocco, 6–8 January 2020; pp. 55–62.
42. Udokwu, C.; Norta, A. Deriving and Formalizing Requirements of Decentralized Applications for Inter-Organizational Collaborations on Blockchain. *Arab. J. Sci. Eng.* **2021**, *46*, 8397–8414. [CrossRef]
43. Norta, A.; Rossar, R.; Parve, M.; Laas-Billson, L. Achieving a High Level of Open Market-Information Symmetry with Decentralised Insurance Marketplaces on Blockchains. In *Intelligent Computing–Proceedings of the Computing Conference, London, UK, 16–17 July 2019*; Springer: Berlin/Heidelberg, Germany, 2019; pp. 299–318.
44. Norta, A.; Hawthorne, D.; Engel, S.L. A Privacy-Protecting Data-Exchange Wallet with Ownership-and Monetization Capabilities. In Proceedings of the 2018 International Joint Conference on Neural Networks (IJCNN), Rio de Janeiro, Brazil, 8–13 July 2018; pp. 1–8.
45. Bamakan, S.M.H.; Motavali, A.; Bondarti, A.B. A Survey of Blockchain Consensus Algorithms Performance Evaluation Criteria. *Expert Syst. Appl.* **2020**, *154*, 113385. [CrossRef]
46. Geiger, M.; Harrer, S.; Lenhard, J.; Wirtz, G. BPMN 2.0: The State of Support and Implementation. *Future Gener. Comput. Syst.* **2018**, *80*, 250–262. [CrossRef]
47. Levy, M. Agent-Based Computational Economics. In *Complex Social and Behavioral Systems: Game Theory and Agent-Based Models*; 2020; pp. 825–849. Available online: https://link.springer.com/referenceworkentry/10.1007%2F978-1-0716-0368-0_6 (accessed on 12 December 2021).
48. Olsen, M. How Firms Overcome Weak International Contract Enforcement: Repeated Interaction, Collective Punishment and Trade Finance. In *IESE Business School Working Paper No. WP-1111-E*; 2016. Available online: https://www.morten-olsen.com/Morten%20Olsen%20-%20how%20firms%20overcome%20weak%20contract.pdf (accessed on 12 December 2021).
49. Timmer, I. Contract Automation: Experiences from Dutch Legal Practice. In *Legal Tech, Smart Contracts and Blockchain*; Springer: Berlin/Heidelberg, Germany, 2019; pp. 147–171.
50. Wood, G. Ethereum: A Secure Decentralized Generalised Transaction Ledger. 2014. Available online: http://gavwood.com/paper.pdf (accessed on 13 October 2021).
51. Buterin, V. A Next-Generation Smart Contract and Decentralized Application Platform–Whitepaper. 2019. Available online: https://github.com/ethereum/wiki/wiki/White-Paper (accessed on 14 October 2021).
52. Wood, G. Polkadot: Vision for a Heterogenous Multi-Chain Framework–Draft 1 (White Paper). 2016. Available online: https://polkadot.network/PolkaDotPaper.pdf (accessed on 8 September 2021).
53. Cap, C.H.; Leiding, B. Blogchain–Disruptives Publizieren auf der Blockchain. *HMD Prax. Wirtsch.* **2018**, *55*, 1326–1340. [CrossRef]

54. Othman, A.; Callahan, J. The Horcrux Protocol: A Method for Decentralized Biometric-based Self-Sovereign Identity. In Proceedings of the 2018 International Joint Conference on Neural Networks (IJCNN), Rio de Janeiro, Brazil, 8–13 July 2018; pp. 1–7.
55. Sporny, M.; Longley, D.; Sabadello, M.; Drummond, R.; Steelie, O.; Allen, C. Decentralized Identifiers (DIDs)-Core Architecture, Data Model, and Representations v1.0. 2021. Available online: https://w3c-ccg.github.io/did-spec/ (accessed on 14 October 2021).
56. Zhou, Q.; Huang, H.; Zheng, Z.; Bian, J. Solutions to Scalability of Blockchain: A Survey. *IEEE Access* **2020**, *8*, 16440–16455. [CrossRef]
57. Dang, H.; Dinh, T.T.A.; Loghin, D.; Chang, E.C.; Lin, Q.; Ooi, B.C. Towards Scaling Blockchain Systems via Sharding. In Proceedings of the 2019 International Conference on Management of Data, Amsterdam, The Netherlands, 30 June–5 July 2019; pp. 123–140.
58. Schär, F. Decentralized Finance: On Blockchain and Smart Contract-based Financial Markets. In *FRB of St. Louis Review*; 2021. Available online: https://papers.ssrn.com/sol3/papers.cfm?abstract_id=3843844 (accessed on 12 December 2021).

Review

Blockchain Technology Applied in IoV Demand Response Management: A Systematic Literature Review

Evgenia Kapassa * and Marinos Themistocleous

Institute for the Future, Department of Digital Innovation, University of Nicosia, Nicosia 2414, Cyprus; themistocleous.m@unic.ac.cy
* Correspondence: kapassa.e@unic.ac.cy

Citation: Kapassa, E.; Themistocleous, M. Blockchain Technology Applied in IoV Demand Response Management: A Systematic Literature Review. *Future Internet* **2022**, *14*, 136. https://doi.org/10.3390/fi14050136

Academic Editors: Ahad ZareRavasan, Taha Mansouri, Michal Krčál, Saeed Rouhani and Paolo Bellavista

Received: 14 March 2022
Accepted: 27 April 2022
Published: 29 April 2022

Publisher's Note: MDPI stays neutral with regard to jurisdictional claims in published maps and institutional affiliations.

Copyright: © 2022 by the authors. Licensee MDPI, Basel, Switzerland. This article is an open access article distributed under the terms and conditions of the Creative Commons Attribution (CC BY) license (https://creativecommons.org/licenses/by/4.0/).

Abstract: Energy management in the Internet of Vehicles (IoV) is becoming more prevalent as the usage of distributed Electric Vehicles (EV) grows. As a result, Demand Response (DR) management has been introduced to achieve efficient energy management in IoV. Through DR management, EV drivers are allowed to adjust their energy consumption and generation based on a variety of parameters, such as cost, driving patterns and driving routes. Nonetheless, research in IoV DR management is still in its early stages, and the implementation of DR schemes faces a number of significant hurdles. Blockchain is used to solve some of them (e.g., incentivization, privacy and security issues, lack of interoperability and high mobility). For instance, blockchain enables the introduction of safe, reliable and decentralized Peer-to-Peer (P2P) energy trading. The combination of blockchain and IoV is a new promising approach to further improve/overcome the aforementioned limitations. However, there is limited literature in Demand Response Management (DRM) schemes designed for IoV. Therefore, there is a need for a systematic literature review (SLR) to collect and critically analyze the existing relevant literature, in an attempt to highlight open issues. Thus, in this article, we conduct a SLR, investigating how blockchain technology assists the area of DRM in IoV. We contribute to the body of knowledge by offering a set of observations and research challenges on blockchain-based DRM in IoV. In doing so, we allow other researchers to focus their work on them, and further contribute to this area.

Keywords: blockchain; smart grid; internet of vehicles; demand response; systematic literature review

1. Introduction

Blockchain was first proposed as Bitcoin's distributed ledger to alleviate the double-spending problem. One of the most important characteristics of blockchain is that, due to its transparency and immutability, it enables participants to establish trust among unknown entities in a decentralized way [1,2]. Recently, in the area of smart mobility, blockchain has risen as an upcoming technology, enabling decentralized mobility services, secure and reliable P2P energy trading between EVs, secure authentication and more [3]. According to Markets and Markets [4], by 2030, the automotive blockchain market is expected to have grown from USD 0.35 billion in 2020 to USD 5.29 billion, drawing the attention of a variety of stakeholders (e.g., investors, business experts, academics and governments).

Additionally, cities are becoming smarter and more connected, due to the rapid advancement of the Internet of Things (IoT). IoT allows connected vehicles (e.g., electric vehicles) to gradually evolve into self-driving vehicles, but none of this will be feasible without a new advanced network [5]. Thus, IoV has emerged as technology that allows vehicle information exchange, efficiency and safety with each other. IoV is powered by smart vehicles, Artificial Intelligence and IoT [6,7]. In the context of IoV, smart vehicles use the Internet to communicate with each other and connect with drivers or passengers, as well as with roadside facilities [8]. The most important communication examples are Vehicle-to-Vehicle (V2V) and Vehicle-to-Grid (V2G) [9].

Yet, the widespread utilization of dispersed EVs creates problems for the energy management in IoV [10,11]. A possible solution is DR management, which might be used in IoV to allow energy consumers (i.e., EV drivers) to adjust their energy consumption patterns based on the cost [12,13]. Specifically, DR allows customers to play an important part in the energy grid operation by decreasing or changing their power usage during peak hours in response to time-based tariffs or other types of financial incentives. Several electric system operators utilize DR programs as alternatives for balancing supply and demand [14]. Such programs can reduce the cost of electricity in energy markets, resulting in reduced market prices. Time-based pricing schemes, including critical peak, variable peak, time of use and real-time pricing, are examples of how market players might participate in demand response [15,16].

Nevertheless, research in the area of IoV DR is still in its early stages and the deployment of IoV DR in smart grids confronts a number of important challenges [17,18]. Specifically, IoV currently lacks sufficient security and privacy procedures to reduce inaccurate and malicious information transfers between EVs [19,20]. There is also a lack of incentive mechanisms to encourage prosumers (i.e., producer and consumer) to join in such DR schemes. EV owners are hesitant to join in large-scale trading networks unless they are highly compensated (i.e., incentivization schemes), due to higher battery drain and other costs associated with discharging [21,22]. Additionally, IoV's characteristics, such as high mobility, low latency, network complexity and heterogeneity, pose substantial issues when typical cloud-based storage and management is incorporated. As a result, to be ready for the future expansion of IoV and fulfill its potential, the data interchange and storage infrastructure may be distributed, decentralized, interoperable, adaptable and scalable [23–26].

From a different point of view, blockchain can be seen as a promising technology in the area of smart mobility and EVs, as it can support secure, reliable and decentralized energy trading [3,27,28]. Blockchain technology provides transparency regarding energy production and consumption. Moreover, blockchain has the potential to give a considerable number of unique solutions in the majority of IoV applications [6]. For that reason, researchers have started developing IoV applications based on blockchain technology [29–31]. IoV built on blockchain has the capacity to boost a new ecosystem for the transportation and vehicular industries, allowing energy assets to be transferred and managed in a secure, transparent, verifiable and efficient manner [32–34].

Despite the fact that IoV is a relatively new technology, it is known that decentralized designs and processes are required to manage energy generation and consumption. Based on our research findings presented in [6], instability in energy production may jeopardize the energy supply security, leading to energy overload and a greatly distributed and continuously changing IoV topology. As a result, we concluded that more study into DR management, which uses blockchain technology to balance energy consumption and supply, was required. In this regard, blockchain appears to have the potential to be an innovative paradigm, addressing limitations in IoV, such as variability in energy production, energy overload and lack of incentive mechanisms [6]. Even though there are several studies that investigate the blockchain technology in the energy domain, there is limited information regarding how blockchain could be incorporated in DR management schemes designed for IoV. Therefore, there is a need to collect and critically analyze the existing literature in an attempt to highlight open issues. Thus, in this article, we conduct a SLR, investigating how blockchain technology assists the area of DRM in IoV. We contribute to the body of knowledge by offering a set of observations and research challenges on blockchain-based DRM in IoV, as the outcome of addressing the following SLR Question: *"How can blockchain technology assist the area of demand response management in IoV-assisted smart grids?"*.

To this end, our study makes the following contributions based on a systematic literature review methodology:

1. Collects and filters the available literature, in an attempt to present current perspectives and research efforts on blockchain-enabled DRM in IoV.

2. Critically analyzes and reports the review's outcomes, in an attempt to discuss the various IoV DRM solutions and scenarios and provide a taxonomy of demand response programs.
3. Focuses on the perspectives and research efforts around the demand response management in the IoV, taking into consideration the application of blockchain technology.
4. Provides a comprehensive list of observations and research challenges of blockchain technology in the IoV DRM.

The rest of this article is structured as follows. Section 2 provides a description of the followed research methodology of the current review, describing step by step the three stages of this articles, based on Kitchenham's approach [35,36], as well as features of the PRISMA methodology [37]. Then, Section 3 presents the main findings and the knowledge extracted during the review around current perspectives and research efforts on blockchain-enabled DRM in IoV, while in Section 4, we critically discuss their findings, providing their main observations, as well as some limitations and research gaps. Finally, Section 5 concludes the paper, highlights its contributions and makes recommendations for further research. Lastly, we would like to mention that we provide a detailed list of abbreviations, in an attempt to ease the understanding of the concepts presented in this article.

2. Systematic Literature Review Methodology

The current study was based on Kitchenham's methodology and on [35,36], as well as on features of the PRISMA statement [37]. This review followed the following steps:

1. Plan the review: Determine the rationale of the review, define the research questions and create the review process.
2. Conduct the review: Carry out the established protocol, select studies and assess their quality.
3. Report the review: Presents the review findings.

2.1. Plan the Review

The current review was built on top of the findings of our previous work presented in [6]. In particular, our earlier research indicated that blockchain offers novel opportunities for vehicle owners to engage in the IoV. However, further research is needed into the DRM. Therefore, continuing our work, the detailed description of the SLR methodology and review protocol can be found in that article. Following that, several components of the current review process were directed by the research questions, including setting inclusion and exclusion criteria, searching for relevant studies, gathering data and presenting findings. The current systematic literature review aims to answer the following main question:

SLR Question: How can blockchain technology assist the area of demand response management in IoV-assisted smart grids?

We identified a set of search queries and databases, while planning the review. It is worth noting that the systematic literature search began in 2017 and spanned the previous five years of research innovation and progress. The search database sources were IEEEXplore, SpringerLink, ScienceDirect, ACM Digital Library and Google Scholar. Because this study focused on the scientific knowledge of blockchain adoption in the IoV concept, we emphasized literature published in academic journals, conference proceedings, and book chapters. Then, we defined the SLR's keywords and the queries that were used in the aforementioned databases (Table 1).

2.2. Conduct the Review

We identified 1254 studies that were related to our criteria. We examined for probable duplication inside the union of all databases' responses before continuing on to the screening procedure. After deleting the duplicates, the results for the screening were 1086, as shown in Figure 1. Moving on to the screening procedure, we assessed the suitability and quality of the collected studies using a set of quality criteria for exclusion and inclusion that we presented in our initial study [6] and present also in Table 2. Then, using those criteria, we looked at the abstracts and keywords of the 1086 papers. During the abstract reading, we concentrated on two qualifying criteria: Is the paper about blockchain? Is there a concept, framework or research in the article that relates to demand response management? To be eligible, papers had to fulfill both requirements. Then, 326 articles were left for review using this method. We continued to evaluate the normative literature after the screening step. We were able to exclude 110 papers because they were unrelated to the review's objective, leaving 106 research to be considered. The approach for identification, screening, and inclusion is depicted in detail in Figure 1.

Table 1. Keywords and search queries for the systematic literature review.

Keyword	Query
blockchain IoV Internet of Vehicles smart grid smart city demand response applications challenges	blockchain AND "demand response" AND (IoV OR "Internet of Vehicles" OR "Smart Grid" OR "Smart City") AND (applications OR challenges)

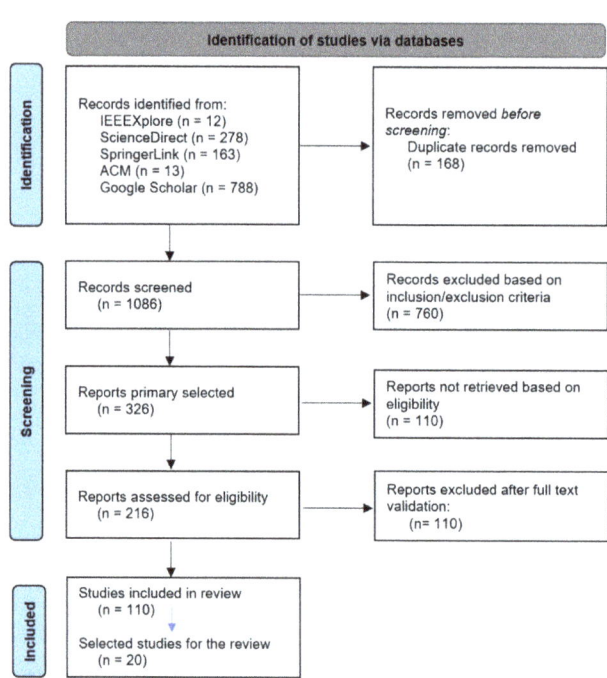

Figure 1. Identification of studies via databases (PRISMA flow diagram).

Following that, the purpose was to use the data extraction to appropriately document the knowledge gathered from the research that were included. The following information was gathered from each study:

- Authors, publication year, paper type, publishing location and digital object identifier were all required fields.
- Evaluation of the study in terms of research knowledge, including the following:

○ The study's issues;
○ The study's results and key findings;
○ The study's limitations and/or research approaches.

A detailed presentation of the research findings and the extracted knowledge is provided in Sections 3 and 4.

Table 2. Inclusion and exclusion criteria.

Inclusion Criteria	Exclusion Criteria
Peer-reviewed studies	Grey literature
Academic theoretical and empirical research	White papers and material from non-academic sources
Full-text available	Full-text not available
Written in the English language	Not written in the English language
Published in 2017 onwards	Published before 2017
Relevant to blockchain and the IoV concept	Diverged from the field of blockchain and the IoV concept
Concept addressed by means of a valid methodology	

2.3. Report the Review

The majority of the studies were journal articles and high-quality conference papers, although some book chapters were also analyzed. A clear depiction of the types of publications identified is presented in Figure 2, while their distribution within the time range of the review is presented in Figure 3. Moreover, their publishers are also presented in Figure 4. From the primary studies (i.e., cluster of 106), the most interesting and relevant ones were selected (excluding the documents in the form of a survey or literature review, which were also considered separately) to highlight the main current research trends and the gaps that have yet to be filled. We selected studies only considering if the paper's argument was built on an appropriate base of theory and concepts, and if the evaluation results were clearly presented and appropriately analyzed. The selected studies were 20 papers and are presented in Appendix A.

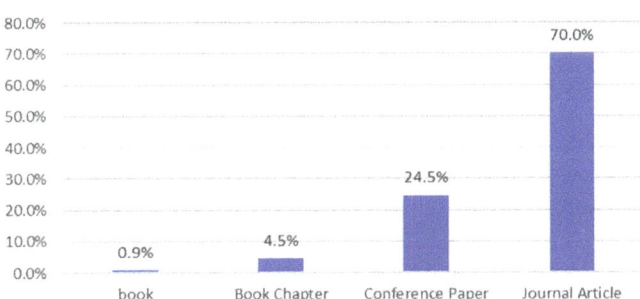

Figure 2. Distribution of the identified studies based on their type.

The following features were highlighted: problem statement, proposed solution/objectives and outcomes/limitations.

Evaluating the selected studies in terms of research knowledge, it appears that current studies partially tackle DRM designed for IoV. There were studies that were separately studied and solved some of the IoV issues, although none succeed in tackling all of them holistically, providing a systematic literature review or focusing on how blockchain could be incorporated into DR management schemes designed for IoV. The aforementioned statement is supported by Table 3, where it can be seen that no work was identified to employ blockchain-based privacy, DRM, V2V/V2G energy trading, charging scheduling, incentivization schemes and EV profiles. Table 3 summarizes and compares the main features observed in the literature review.

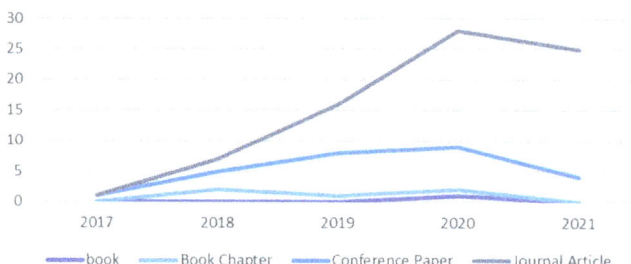

Figure 3. Distribution of the identified studies within the review's time range.

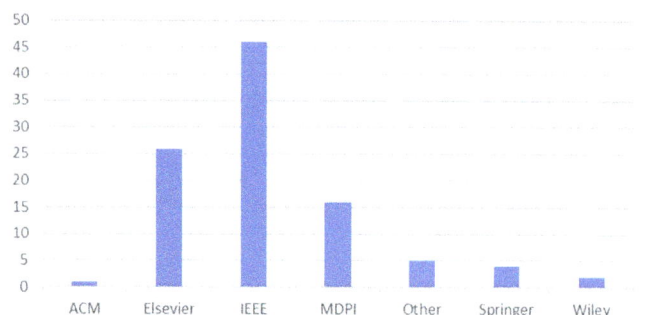

Figure 4. Publishers for the identified primary studies.

Table 3. Inclusion and exclusion criteria.

Reference (Selected Studies)	Blockchain-Based Privacy	Demand Response Management	V2V/V2G Energy Trading	Charging Scheduling	Incentive Mechanism	EV Profiles
[38]	•		•	•		
[39]	•		•			
[40]		•				
[41]		•		•	•	
[42]				•		•
[43]	•		•			
[44]			•			
[45]	•	•	•			
[46]			•	•		
[47]		•			•	
[48]					•	•
[12]	•	•	•			
[49]	•		•	•	•	
[50]					•	
[51]		•				•
[52]	•		•			
[53]	•					
[54]	•			•		
[55]		•	•			•
[56]	•	•			•	
This SLR	•	•	•	•	•	•

3. Current Perspectives and Research Efforts on Blockchain-Enabled IoV

This section summarizes the research findings derived from the systematic literature review, focusing on the perspectives and research efforts around the demand response management in the IoV, taking into consideration the application of blockchain technology.

3.1. P2P Trading and Management in Energy Blockchain

There are several studies that propose the use of blockchain technology in P2P energy trading [57]. For instance, in [44], a game-theoretic approach for a demand side management model that incorporates a localized PBFT-CB was proposed. The study incorporated the interaction between sellers and buyers, considering the Stackelberg game and non-cooperative static games. Additionally, Brooklyn microgrid was one of the first applied engineering programs of energy blockchain [58]. The project is based on blockchain P2P energy trading without the intermediation of a third-party energy supplier. The Brooklyn microgrid demonstrates that blockchain may be utilized in real-world P2P energy transactions. Moreover, Q. Duan [59] proposed an optimal scheduling and management smart city scheme, within the safe framework of blockchain. To do so, Q. Duan presented an enhanced directed acyclic graph strategy to increase the security of data transactions inside a smart city, as well as a security layer based on blockchain to prevent cyber hacking. Additionally, the LO3 Energy company introduced an energy supply scheme to the closest neighbors based on P2P trading [60]. Lastly, the current literature dictates future distributed ledger implementations and mechanisms and revealed that blockchain is an important part of P2P energy trading [52,61–63].

3.2. Blockchain-Based Demand Response Programs and Optimization Models

Demand response has been recognized as an important tool for managing supply and demand in electrical grids [64]. When there is an electrical deficit, DR becomes an effective alternative for absorbing the energy gap and managing power utilization [47]. Significant initiatives in blockchain-based demand response programs and optimization models are presented below.

To handle demand response in a V2G context, a P2P energy trading mechanism between EVs and network operators was proposed by S. Aggarwal [39], in an attempt to overcome smart grid imbalances and to control the ever-growing energy demands from EVs. Moreover, Z. Guo [47] presented a blockchain-enabled DR scheme with an incentive pricing model. First, the authors proposed a blockchain-enabled DR framework to promote the secure implementation of DR, while then they also designed a dual-incentive mechanism, based on the Stackelberg game model, to successfully implement blockchain demand response management. Furthermore, in the area of IoV, there are also several studies that address the DR problem and propose optimization solutions. For instance, Z. Zhou [54] proposed a consortium blockchain-enabled secure energy trading framework for EVs with a moderate cost, using a contract theory-based incentive mechanism to incentivize more EVs to participate in DR. The proposed optimization scheme falls into the category of difference of convex programing and is solved by using the iterative convex–concave procedure algorithm. Likewise, T. Zhang [41] incorporated a blockchain-based cryptocurrency component, with which the system can incentivize users with monetary and non-monetary means in a flat-rate manner.

In recent years, the implementation of DR programs in smart grids has drawn a lot of academic attention. A taxonomy of these research endeavors is depicted in Figure 5, which was generated from the outcomes of the current SLR. This categorization is based on the DR procedure's control mechanism, customer motives to lower or move their expectations and the DR decision variable.

DR systems have two types of control mechanisms: centralized and distributed. In the centralized mode, consumers connect directly with the electricity network without engaging with one another. In the distributed mode, user interactions feed the network with information about overall usage [65].

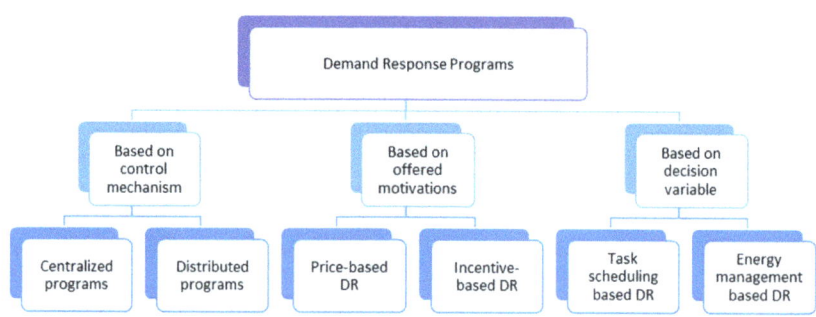

Figure 5. Demand response program taxonomy.

The motivations offered to producers and consumers to decrease their energy usage are classified in the second category of DR schemes. These motivations are divided into two categories:

- Time-based DR: In the time-based DR, consumers are provided time-varying pricing depending on the cost across various time periods.
- Incentive-based DR: Customers in incentive-based DR schemes are offered fixed or time-varying payments to encourage them to reduce their electricity usage during times of system stress [66–68], but they are also subject to specific constraints or are penalized if they do not participate in the program.
- Finally, DR systems that utilize the decision variable to identify task-scheduling and energy-management based DR schemes are classified into the third group [69,70].

The main feature of task scheduling DR is control over the desired load's activation time, which may be moved to peak-demand periods [71,72]. The energy-management-based DR solutions accomplish different power usage during peak-demand hours by decreasing the power consumption of certain loads [73,74].

3.3. Electric Vehicles Charging Scheduling Using Blockchain

To study the consequences of increased EV load and charging mechanisms, the accurate modeling of EV charging profiles is necessary [75]. The size and topology of the energy grid, the number and size of EVs, the mode, time and location of charging as well as the daily driving distance influence the above-mentioned charging profiles. As a consequence, the charging profiles of the drivers are increasingly coupled with charging schedules.

N. Guo et al. [76] proposed a centralized control architecture to handle the modeling and management of EV charging by reducing peak demand and increasing the number of EVs charged concurrently. A common finding in numerous studies on EV battery chargers is that EV battery workloads are commonly thought of as a static, with the actual system behavior of the batteries throughout the charging process being overlooked. In order to tackle the latter, Y. Wu [77] emphasized that a bi-directional energy flow is conceivable. EV batteries may be utilized in the grid in the manner of any other energy storage device, with the additional perk of mobility. The owners of EVs would be able to participate in energy market trading, recharge batteries when energy is cheap and discharge if the smart grid rewards them for their excess energy. This type of energy exchange and negotiation can allow the network to regulate demand (e.g., peak shaving) or offer additional storage in the case of excess renewable energy generation. Consumers will be able to pick where, when and which EV to charge, reducing the strain of the grid.

Using the adaptable EV charging flow, C. Lazaroiu [42] developed a model for smart charging of EVs, in which a software agent selects whether it should load a unit, in what sequence or whether it is better to sell energy to the market. The concept is based on blockchain technology, which makes interactions reliable and traceable, with the goal of decreasing or eliminating intermediaries in energy trade and lowering anxiety.

Furthermore, given the growing popularity of EVs and their unpredictable dynamic nature in terms of charging and route patterns, EV load might be difficult for energy distribution operators and utilities to manage [41]. Thus, T. Zhang proposed SMERCOIN, which is a real-time solution that integrates the concepts of priority and cryptocurrencies to encourage EV owners to charge on a renewable energy-friendly timeframe. Customers with a longer history of utilizing renewable energy are given priority in the system, which uses a rating system. By including cryptocurrency, the system may encourage users using both monetary and non-monetary techniques in a flat-rate manner. Similarly, Z. Zhou [54] developed a distributed, privacy-preserved and incentive-compatible DR mechanism for IoV. In more detail, the authors suggested a low-cost consortium blockchain-enabled secure energy trade platform EVs, as well as an incentive system based on contract theory in order to encourage more EVs to join the DR program.

Furthermore, a dependable solution is required to meet the future energy demands of urban and industrial customers, while also supporting the charging and discharging needs of EVs. Therefore, some research articles have been presented, studying the energy trading in the IoV for demand response management. For example, S. Aggarwal [39] proposed a blockchain-based secure energy trading scheme for demand response management between EVs and the service providers, while a double auction mechanism is proposed between EVs and SPs to maximizes social welfare with privacy preservation. Similar examples can also be found in [12,46], in which the authors analyzed the energy in IoV-assisted smart cities, employing blockchain capabilities in order to select the most suitable charging station without sharing private information, and to balance the spatio-temporal dynamic demands of computing resource.

4. Discussion

The majority of the selected studies mention the need for decentralized DR management in IoV and smart grids, which will primarily promote privacy and security, and then will efficiently incorporate the increasing number of electric vehicles. Likewise, the need for a blockchain framework that offers optimized management and coordination of EV charging is another similarity identified in the literature. The latter is further supported from the common belief that blockchain technology could provide security and privacy of the drivers' data and the exchange of information. In addition, many of the studies highlight that focus should be given to human behavior and preferences, creating EV profiles, which will help the management of the IoV-assisted smart grids and also emphasize different social aspects. Furthermore, another similarity revealed that real-time energy demand is not sufficiently analyzed and explored, considering also the randomness of the EVs events and the unexpected events that may occur during the everyday life.

Finally, the majority of the studies highlight that the advantages of blockchain technology within the IoV-assisted smart grids are numerous, although there is a lack of incentivization schemes that provide relevant rewards to prosumers in order to participate in such schemes. The similarities are depicted in Table 4.

Table 4. Similarities and differences among the selected studies.

Category	Similarities	Differences
Incentivization	Blockchain incentives are needed to encourage participation	Focus on Non-Fungible Tokens (NFTs) as a mean for incentivization scheme
Privacy and Security	Blockchain technology is mostly used for security and privacy	Data analytics scheme for security-aware DRM using blockchain
Demand Response Management	Real-time demand management is not investigated	Incorporate deep learning for intelligent demand response
EV drivers' profile	Drivers' preferences are not considered	n/a
Generic	Consortium blockchain is common in the DRM applications	The proposition is not directly applied in EVs

Furthermore, the SLR reveled some differences (Table 4). The first difference derives from Y.C. Tsao [40], in which the sustainable microgrid design problem is addressed by leveraging blockchain technology to provide real-time-based demand response programs. The study, though, is not coupled with IoV or EVs, although the optimization approach that is proposed is evaluated, making the authors believe that it could be also applied in the area of IoV. Additionally, compared with the rest of the selected studies that are focused on price-based incentives, N. Karandikar [50] focused on non-fungible tokens as a mean for the incentivization of the users. On the contrary, B. Prapadevi [45] reviewed four important themes, such as electric load forecasting, state estimation, energy theft detection and energy sharing and trading, trying to illustrate the need of deep learning solutions in smart grids and demand response. Similarly, A. Kumari [56] reported that current DR management solutions are not adequate in terms of peak loads reduction, consumer comfort and data security issues, and proposed a data analytics scheme for security-aware management. The proposed scheme used blockchain to maintain the grid stability and reduce peak energy consumption.

The discussion presented above led to the following observations, as those are illustrated in Figure 6.

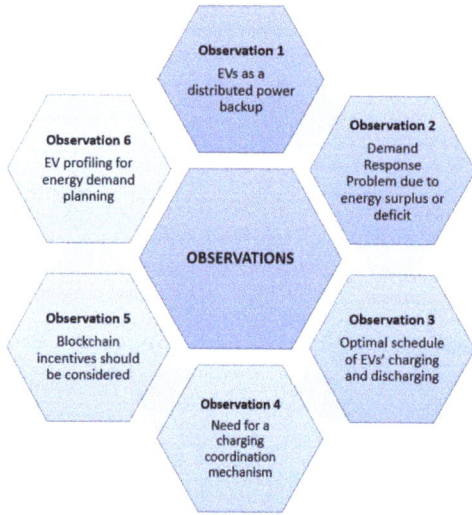

Figure 6. Systematic literature review observations.

Observation 1—EVs as a distributed power backup: EVs can be used as a distributed backup power for the grid, storing electricity during the low period and providing electricity to the power grid during peak period. EVs are not only charged from the grid, but they can also discharge electricity towards that through V2G technology. Hence, EVs may be considered as a distributed backup power, enabling electricity storage during low demand periods, while providing electricity back to the grid during peak periods.

Observation 2—Demand Response Problem due to energy surplus or deficit: Energy surplus or deficit may threaten the security of the energy supply and demand, leading to a demand response problem. The latter is becoming worse considering the randomness of the EV events, which may lead to energy components' overload and culminating with power outages or service disruptions, leading to the so-called demand response problem.

Observation 3—Optimal schedule of EVs' charging and discharging: It is challenging to optimally schedule the charging/discharging behavior of EVs to achieve energy balance, considering the instability of EV demand during specific time periods and/or locations. Specific areas of the IoV-assisted smart grid may increase the demand during specific time periods and/or locations. Thus, it is a challenge to optimally schedule the charging/discharging behavior of EVs in order to achieve energy balance in the grid.

Observation 4—Need for a charging coordination mechanism: Existing pricing coordination techniques have a number of flaws since they rely on a single entity, which might be an untrustworthy third party who is not always truthful when scheduling charging requests. Furthermore, private information about EV owners (such as driving patterns and profiles) could be revealed. To tackle the DR problem in IoV-assisted smart grids, a decentralized, transparent, and privacy-preserving charging coordination mechanism is required.

Observation 5—Blockchain incentives should be considered: Most of the studies consider blockchain technology to ensure the privacy of the EVs, although few of them consider incentive mechanisms to encourage EV drivers to participate in blockchain-enabled DR through optimal scheduling. Existing charging coordination mechanisms suffer from several limitations, e.g., they rely on a single entity, which may be an untrusted party that is not always honest in scheduling charging requests. In most of the selected studies, it was observed that the researchers considered blockchain technology to ensure the privacy of EV owners. However, not many of them considered incentive mechanisms to encourage EV drivers to participate in this kind of blockchain-enabled DR framework. There are a couple of studies, though, that state that the provision of incentives to the participants (e.g., EV drivers, energy providers and households) will be the key to exploit blockchain technology within smart grids and IoV.

Observation 6—EV profiling for energy demand planning: EV profiling should be considered to perform an alignment of EV charging and driver mobility demand towards optimizing electricity demand forecasting and planning. Forecasting the electricity price plays a significant role in reducing energy costs. Moreover, energy demand forecasting helps to maintain the balance between electricity demand and supply in the IoV-assisted smart grid. As a consequence, to achieve the optimization of electricity demand forecasting and planning, EV profiling should be considered.

As discussed in earlier sections, IoV is now under strain as a result of substantial changes in the production and development of EVs. Indeed, the growing malfunctions in power generation need the development of new paradigms. Demand response is an approach in which EV customers actively alter their consumption in response to grid demands. Thus, energy management, which allows the optimal use of constrained energy resources, is required for the establishment of a smart, green and sustainable smart grid. However, the widespread use of unpredictable and uncoordinated EVs creates problems. To balance load and supply, a large number of centralized generators and energy storage devices should be placed, resulting in a considerable CAPEX and operational expense OPEX. Another option is to investigate the rapid spread of DR, which may be used in smart cities to allow energy users to proactively change how and when they use (or create) energy based on the cost (or reward). Because IoV is a participatory data exchange and storage platform, the underlying information exchange system has to be safe, transparent and immutable in order to accomplish the desired objectives. In this regard, the use of blockchain as a system platform for addressing the IoV's demands was investigated. IoV applications enabled by blockchain are thought to offer a variety of desirable features, such as decentralization, security, transparency, immutability and automation, due to their decentralized and immutable nature.

Even though the current studies have several similarities, there are still open research challenges that need to be further investigated. The identified challenges are described below among with some suggestions for further investigation.

Research Challenges and Suggestions 1:

- Research Challenge 1: EV information is exposed, resulting in privacy and security issues.
- Suggestion 1: Blockchain infrastructure and identity management for secure information exchange in IoV.

- Description: The existing charging coordination mechanisms suffer from their relation to a single entity (e.g., the charging coordinator), which can reveal private information about the owners of the EVs (e.g., patterns and drivers' profiles). Thus, the integration of blockchain in the IoV should guarantee the privacy of all participants and the security of the exchanged information.

Research Challenges and Suggestions 2:

- Research Challenge 2: Demand and response in IoV are affected by energy generation and consumption.
- Suggestion 2: V2V/V2G Energy Trading considering EVs' Charging Scheduling addressing the Demand Response Problem.
- Description: The widespread use of unpredictable dispersed RES and uncoordinated EVs creates problems for smart energy management. Current studies are investigating the optimization of the charging scheduling of EVs, although they do not consider the regional energy balance, leading to demand–response gaps and energy imbalances. Thus, emphasis should be given to the energy demand and response of the EVs in specific regions of a smart grid (e.g., considering social events and/or accidents).

Research Challenges and Suggestions 3:

- Research Challenge 3: EV charging profiling from an EV user perspective is not investigated.
- Suggestion 3: EV profiling for optimal charging scheduling and DR balance.
- Description: EV charging profiling from an EV user perspective is not sufficiently investigated. This means that each EV user should be aware of and declare its charging preferences and also to update this information in a continuous manner. In order to successfully control the charging/discharging schedule in comparison to IoV metrics and stability, a certain amount of smartness should be considered.

Research Challenges and Suggestions 4:

- Research Challenge 4: Due to a lack of incentives, EVs with excess energy are not encouraged to act as energy marketers.
- Suggestion 4: Incentive provisioning through rewards and penalties.
- Description: There is too little work conducted in the area of incentivization mechanisms. The majority of the studies do not consider any incentive mechanism to encourage EV drivers to participate in a blockchain-enabled DRM scheme. Therefore, it is necessary to provide an effective incentivization scheme that will give the appropriate rewards and/or penalties to the IoV participants and exploit the blockchain related activities.

5. Conclusions

The current review explores the application of blockchain technology in the rising concept of IoV demand response management, investigating in a systematic way the literature from the beginning of 2017 until the end of and 2021. We satisfied the goals of this review and answered the following research questions: (a) How does blockchain promote the P2P trading among EVs? (b) What is the current status on blockchain-based demand response programs and optimization models and which are the most common techniques for demand response management? and (c) What research work has been conducted regarding EV charging scheduling using blockchain? It is worth mentioning that we extracted knowledge following a systematic methodology based on Kitchenham's approach and present their findings around current perspectives and research efforts on blockchain-enabled IoV DR management. Although we found a vast number of papers using a thorough search procedure, some of them were judged irrelevant. Our findings are categorized in three parts related to: (a) P2P trading and management in energy blockchain, (b) blockchain-based demand response programs and optimization models and (c) electric vehicle charging scheduling using blockchain. Finally, the current study concludes by providing the outcomes of the systematic literature review, highlighting our main observations and opening research challenges. Additionally, we provided an analysis of the similarities and differences between the reviewed articles, showing, at the end, a set of limitations in the literature. This work goes beyond the currently available studies that focus on the blockchain application in the energy domain in general. Rather than that, the novelty of this study lies in the fact that it provides a systematic literature review in the area of DRM in IoV, based on blockchain technology. Therefore, it provides a thorough analysis of specific parts of the energy domain, emphasizing the above-mentioned perspectives and research efforts on blockchain-enabled IoV.

Our key takeaway from this study is that the disruption of blockchain in IoV is increasing at a fast pace. Currently, there are some studies that tackle the identified research challenges, although none tries to solve them holistically, as it should require a real-world scenario. Thus, we plan to extend the current review to propose more detailed solutions to overcome the identified research challenges. In that sense, we plan to further investigate the need of a unified blockchain framework that tackles all the identified challenges in a holistic way, considering secure energy trading, optimal charging scheduling and motivation towards the demand response management. It is also believed that further research is needed in the area of blockchain-enabled IoV to exploit its full potential and understand the limitations when applied in large-scale deployments.

Author Contributions: Conceptualization, methodology, validation, formal analysis, investigation, data curation, visualization, E.K.; writing—original draft preparation, E.K.; writing—review and editing, E.K. and M.T.; supervision, M.T. All authors have read and agreed to the published version of the manuscript.

Funding: This work was supported by PARITY project funded by the European Union's Horizon 2020 Framework Program for Research and Innovation under grant agreement no. 864319.

Institutional Review Board Statement: Not applicable.

Informed Consent Statement: Not applicable.

Data Availability Statement: Not applicable.

Acknowledgments: This work was supported by the Institute for the Future (IFF), University of Nicosia, as part of the corresponding author's (E.K.) Ph.D. studies.

Conflicts of Interest: The authors declare no conflict of interest.

Abbreviations

DER	Distributed Energy Recourse
DR	Demand Response
DRM	Demand Response Management
DRP	Demand Response Problem
EV	Electric Vehicle
IoT	Internet of Things
IoV	Internet of Vehicles
ITS	Intelligent Transportation Systems
P2P	Peer to Peer
RES	Renewable Energy Sources
RSU	Road Side Unit
SLR	Systematic Literature Review
V2G	Vehicle-to-Grid
V2V	Vehicle-to-Vehicle
V2X	Vehicle-to-Everything

Appendix A

Table A1. Selected Studies.

Title	Problem Description	Study Outcomes/Objectives	Limitations
[38]	Different prices based on demand and response, privacy issues and detection of customers' and EVs' position.	A reliable, automated and privacy-preserving selection of charging stations based on pricing and distance to the electric vehicle.	Possibility of denial-of-service attack. Charging stations are not fully utilized or EVs are not guaranteed a time slot.
[39]	Uncoordinated usage and unregulated energy demand from EVs may increase the demand–supply gap between the service providers and the consumers.	A Peer-to-Peer (P2P) energy trading scheme between EVs and the SPs to manage the demand response in V2G environment, providing incentives to EVs. Consortium blockchain is used to ensure secure energy transactions between EVs and the SPs without a trusted third-party intervention.	Energy scheduling is not considered; Optimal EV charging is not considered
[40]	Sustainable microgrids that simultaneously address economic benefits, environmental and social issues have not been broadly explored by researchers.	Leveraging blockchain technology to provide real-time-based demand response programs.	Blockchain-based smart contracts should be considered in sustainable microgrids to ensure a fair deal for various stakeholders.
[41]	Random dynamic nature of electric vehicle charging and routing cause issues in the electric vehicles' load and could challenge the power distribution operators and utilities.	A real-time system that incorporates the concepts of prioritization and cryptocurrency to incentivize electric vehicle users to collectively charge with a renewable energy-friendly schedule. The study incorporated a blockchain-based cryptocurrency component in order to incentivize users with monetary and non-monetary means in a flat-rate system.	The study was designed based on a photovoltaic generation system and is not evaluated in IoV scenarios.
[42]	The rising demand for electric vehicles will necessitate an increase in charging infrastructures, both to ensure charging system absorption and to disperse energy demand.	A blockchain-based approach for smart charging of electric vehicles, in which a software agent determines whether to load a machine, in what order or whether it is preferable to sell energy to the retail market. The agent adjusts to the individual prosumers of electric vehicles, learning their preferences and mobility habits, so that owners of electric vehicles choose to participate in the system.	Real-time demand is not addressed and blockchain incentives are not clear enough.
[43]	Demand response procedures are transmitted in the smart city with the use of communication infrastructures, which can lead to a variety of attacks in which a malicious user can exploit security flaws in the network.	A safe demand response management system based on blockchain that secures energy trade choices for controlling the total load of domestic, commercial and industrial sectors.	The latency of the proposed system should be decreased and throughput should be increased. Incentives are not present.
[44]	While electricity trading plays an important role in P2P trading, the existing studies have not analyzed the interaction among prosumers regarding pricing.	A game-theory-based pricing model in PBFT-based consortium blockchain is proposed, as well as a rule-based iterative pricing algorithm to obtain the equilibrium prices.	Energy profiles are not taken into consideration neither scheduling algorithms are in place.

Table A1. *Cont.*

Title	Problem Description	Study Outcomes/Objectives	Limitations
[45]	A large amount of data are generated every day in demand response systems from different sources, such as energy production (e.g., wind turbines), transmission and distribution (e.g., microgrids) and load management (e.g., smart meters and electric vehicles).	Analysis of deep learning applications in smart grids and demand response, including electric load forecasting, state estimation, energy theft detection and energy sharing and trading.	Aspects such as dynamic pricing for demand response. load forecasting in smart grids and EV scheduling are not discussed.
[46]	The untrustworthy centralized nature of energy markets and EV charging infrastructures expose EV users' personal information to a number of privacy and security risks.	A blockchain-based charging station selection mechanism for electric vehicles, that ensures EV users' confidentiality and privacy, availability of reserved time slots at the charging stations, Quality of Service (QoS) and improved EV user comfort.	The use of dynamic pricing is restricted. Although it is a vital part in unleashing EVs' flexibility potential, which is necessary for the future grid integration of EVs and renewable energy.
[47]	Demand response necessitates the use of a central agent, which raises security and trust concerns. Furthermore, during incentive pricing, disparities in user response cost features are not considered, affecting the equitable participation of users in DR and increasing expenses.	A blockchain-enabled demand response scheme with an individualized incentive pricing mode is proposed.	More market-realistic scenarios, such as more than one power retail firm engaging in demand response and a higher number of consumers, must be considered. Investigate game and solution models that are appropriate for market-realistic scenarios.
[48]	Increasing available supply to match the projected peak usage value requires the energy operator to over-provision the generation capacity, which can be expensive.	A blockchain-based and data-driven approach for incentive-based peak mitigation.	The study was not implemented in the context of IoV. Additionally, real-time re-scheduling based on unforeseen events was not considered.
[12]	Due to their selfishness and mistrust, smart vehicles with excessive computational power may be hesitant to join in the trading process.	To ensure transaction security and anonymity, a consortium blockchain approach is used. The authors used a consortium blockchain approach to show how to trade safe computing resources and entice individual smart automobiles to join the system.	Energy scheduling is not sufficiently analyzed, neither sufficient incentives are provided to participate in the blockchain demand response network.
[49]	Heterogeneous entities on the demand side pose a risk to the power system's reliability and security.	For demand side management, a blockchain-enhanced price incentive demand response is presented. Data verification is recommended to check the validity of the data completed by each user, based on blockchain capabilities, to ensure the credibility of the best energy schedule. All users retain data that are visible, traceable and tamper-proof.	Energy scheduling is not sufficiently analyzed.
[50]	Peak demand times provide a problem to the grid operator since they may need over-provisioning the grid capacity in order to preserve system stability, raising the marginal cost of energy.	Present a unified blockchain-based energy asset transaction system for prosumers, electric cars, power companies and storage providers, incorporating fungible and non-fungible tokens.	Focusing on token incentives, but not on the demand scheduling.

Table A1. Cont.

Title	Problem Description	Study Outcomes/Objectives	Limitations
[51]	Because centralized approaches in smart grid management are no longer effective, the necessity for innovative decentralized techniques and designs are generally acknowledged.	A distributed ledger storage and management solution based on blockchain for energy data gathering from IoT and smart metering devices. Self-enforcing smart contracts are also proposed for programmatically specifying the expected energy flexibility at the prosumer level, the related incentives or penalties and the rules for balancing energy demand with energy output at the grid level.	It was pointed that currently the Distributed System Operator is still on control in a centralized manner.
[52]	There are several challenges that consumers and smart grids face when it comes to user's data, including traceability, authorization, data integrity, data security and single point of failure.	The decentralized nature of the local market is highlighted by the usage of a distributed blockchain technology. Through the Periodic Double Auction method, the study provides a decentralized market platform for trading locally without the need for a central middleman.	Decentralized storage is not present.
[53]	Demand response program acceptance is still lacking owing to consumers' lack of understanding, fear of losing control and privacy over their energy data, and other factors.	A decentralized solution for demand response programs on top of a public blockchain that uses zero-knowledge proofs to protect the privacy of the prosumer's energy data and uses smart contracts to validate the prosumer's behavior inside the program on the blockchain.	Smart grid services have varying response time requirements, which affects the accuracy required for energy data monitoring and the costs of integrating an energy blockchain.
[54]	Internet of electric vehicles lacks incentive mechanism and suffers from privacy leakage and security threats.	A blockchain-enabled safe energy trading system for privacy and security in the Internet of vehicles.	Given that the data in a block are encrypted using asymmetric encryption techniques, decrypting them without knowing the secret key is extremely expensive. The computation resources required to determine a block are prohibitive, preventing the widespread adoption of blockchain-based energy trade.
[55]	The extensive deployment of EVs can bring challenges to the grid if not properly integrated.	Propose blockchain-based smart contracts that allow decentralized energy trading among EVs, considering the users' preferences for the charging scheduling models.	Real-time rescheduling of the charging procedure is not considered.
[56]	Increased demand–response gaps and poor service quality of contemporary ICT-based smart grid in industry 4.0 are caused by the exponential rise in energy demand, necessitating the urgent need for an effective Demand Response Management system to address the aforementioned issues. In terms of peak load reduction, customer satisfaction and data security concerns, the available options are insufficient.	A Demand Response Management algorithm is suggested, combined with a customer incentive system, to minimize peak energy usage. The authors propose an Ethereum-based smart contract to address security concerns and the InterPlanetary File System (IPFS) to address data storage costs.	Dynamic pricing strategies, as well as real-time rescheduling concerns, should be explored.

References

1. Vinet, L.; Zhedanov, A. A "Missing" Family of Classical Orthogonal Polynomials. *J. Phys. A Math. Theor.* **2011**, *44*. [CrossRef]
2. Kapassa, E.; Themistocleous, M.; Quintanilla, J.R.; Touloupos, M.; Papadaki, M. Blockchain in Smart Energy Grids: A Market Analysis. In *Lecture Notes in Business Information Processing*; Springer International Publishing: New York, NY, USA, 2020; Volume 402, pp. 113–124. [CrossRef]
3. Jabbar, R.; Dhib, E.; Said, A.B.; Krichen, M.; Fetais, N.; Zaidan, E.; Barkaoui, K. Blockchain Technology for Intelligent Transportation Systems: A Systematic Literature Review. *IEEE Access* **2022**, *10*, 20995–21031. [CrossRef]
4. MarketAndMarkets Automotive Blockchain Market. Available online: https://www.marketsandmarkets.com/Market-Reports/automotive-blockchain-market-150652065.html (accessed on 10 March 2022).
5. Salem, A.H.; Damaj, I.W.; Mouftah, H.T. Vehicle as a Computational Resource: Optimizing Quality of Experience for connected vehicles in a smart city. *Veh. Commun.* **2022**, *33*, 100432. [CrossRef]
6. Kapassa, E.; Themistocleous, M.; Christodoulou, K.; Iosif, E. Blockchain Application in Internet of Vehicles: Challenges, Contributions and Current Limitations. *Future Internet* **2021**, *13*, 313. [CrossRef]
7. Xin, Q.; Alazab, M.; González Crespo, R.; Enrique Montenegro-Marin, C. AI-based quality of service optimization for multimedia transmission on Internet of Vehicles (IoV) systems. *Sustain. Energy Technol. Assess.* **2022**, *52*, 102055. [CrossRef]
8. Rasheed Lone, F.; Kumar Verma, H.; Pal Sharma, K. Evolution of VANETS to IoV. *Teh. Glas.* **2021**, *15*, 143–149. [CrossRef]
9. Mahmood, Z. Connected vehicles in the IoV: Concepts, technologies and architectures. In *Connected Vehicles in the Internet of Things: Concepts, Technologies and Frameworks for the IoV*; Mahmood, Z., Ed.; Springer International Publishing: Cham, Switzerland, 2020; pp. 3–18. ISBN 9783030361679.
10. Kadhim, A.J.; Naser, J.I. Toward Electrical Vehicular Ad Hoc Networks: E-VANET. *J. Electr. Eng. Technol.* **2021**, *16*, 1667–1683. [CrossRef]
11. Manzolli, J.A.; Trovão, J.P.; Antunes, C.H. A review of electric bus vehicles research topics–Methods and trends. *Renew. Sustain. Energy Rev.* **2022**, *159*, 112211. [CrossRef]
12. Lin, X.; Wu, J.; Mumtaz, S.; Garg, S.; Li, J.; Guizani, M. Blockchain-based On-Demand Computing Resource Trading in IoV-Assisted Smart City. *IEEE Trans. Emerg. Top. Comput.* **2020**, *9*, 1373–1385. [CrossRef]
13. Maximilian, J. Blaschke Dynamic pricing of electricity: Enabling demand response in domestic households. *Energy Policy* **2022**, *164*, 112878.
14. Verbič, G.; Mhanna, S.; Chapman, A.C. Energizing Demand Side Participation. In *Pathways to a Smarter Power System*; Taşcıkaraoğlu, A., Erdinç, O., Eds.; Academic Press: New York, NY, USA, 2019; pp. 115–181. ISBN 978-0-08-102592-5.
15. Venkatachary, S.K.; Prasad, J.; Samikannu, R. Challenges, opportunities and profitability in virtual power plant business models in Sub Saharan Africa-Botswana. *Int. J. Energy Econ. Policy* **2017**, *7*, 48–58.
16. Guo, B.; Weeks, M. Dynamic tariffs, demand response, and regulation in retail electricity markets. *Energy Econ.* **2022**, *106*, 105774. [CrossRef]
17. Aggarwal, S.; Chaudhary, R.; Aujla, G.S.; Kumar, N.; Choo, K.K.R.; Zomaya, A.Y. Blockchain for smart communities: Applications, challenges and opportunities. *J. Netw. Comput. Appl.* **2019**, *144*, 13–48. [CrossRef]
18. Akhtar, N.; Patil, V. *Electric Vehicle Technology: Trends and Challenges*; Springer: Berlin, Germany, 2022.
19. Gill, S.S.; Tuli, S.; Xu, M.; Singh, I.; Singh, K.V.; Lindsay, D.; Tuli, S.; Smirnova, D.; Singh, M.; Jain, U.; et al. Transformative effects of IoT, Blockchain and Artificial Intelligence on cloud computing: Evolution, vision, trends and open challenges. *Internet Things* **2019**, *8*, 100118. [CrossRef]
20. Akhter, A.F.; Ahmed, M.; Shah, A.F.; Anwar, A.; Kayes, A.S.; Zengin, A. A blockchain-based authentication protocol for cooperative vehicular ad hoc network. *Sensors* **2021**, *21*, 1273. [CrossRef]
21. Moniruzzaman, M.; Yassine, A.; Benlamri, R. Blockchain-based Mechanisms for Local Energy Trading in Smart Grids. In Proceedings of the HONET-ICT 2019-IEEE 16th International Conference on Smart Cities: Improving Quality of Life using ICT, IoT and AI, Charlotte, NC, USA, 6–9 October 2019; pp. 110–114.
22. Visakh, A.; Parvathy, S.M. Energy-cost minimization with dynamic smart charging of electric vehicles and the analysis of its impact on distribution-system operation. *Electr. Eng.* **2022**. [CrossRef]
23. Faruk, M.J.H.; Shahriar, H.; Valero, M.; Sneha, S.; Ahamed, S.I.; Rahman, M. Towards Blockchain-Based Secure Data Management for Remote Patient Monitoring. In Proceedings of the 2021 IEEE International Conference on Digital Health, ICDH 2021, Chicago, IL, USA, 5–10 September 2021; pp. 299–308.
24. Latif, S.; Idrees, Z.; e Huma, Z.; Ahmad, J. Blockchain technology for the industrial Internet of Things: A comprehensive survey on security challenges, architectures, applications, and future research directions. *Trans. Emerg. Telecommun. Technol.* **2021**, *32*, e4337. [CrossRef]
25. Ye, X.; Li, M.; Yu, F.R.; Si, P.; Wang, Z.; Zhang, Y. MEC and Blockchain-Enabled Energy-Efficient Internet of Vehicles Based on A3C Approach. In Proceedings of the 2021 IEEE Global Communications Conference (GLOBECOM), Madrid, Spain, 7–11 December 2021; pp. 1–6.
26. Zhang, L.; Cheng, L.; Alsokhiry, F.; Mohamed, M.A. A Novel Stochastic Blockchain-Based Energy Management in Smart Cities Using V2S and V2G. *IEEE Trans. Intell. Transp. Syst.* **2022**, 1–8. [CrossRef]
27. Fotiou, N.; Pittaras, I.; Siris, V.A.; Voulgaris, S.; Polyzos, G.C. Secure IoT access at scale using blockchains and smart contracts. In Proceedings of the 20th IEEE International Symposium on A World of Wireless, Mobile and Multimedia Networks, WoWMoM, Washington, DC, USA, 10–12 June 2019; Institute of Electrical and Electronics Engineers Inc.: Piscataway, NJ, USA, 2019.

28. Asfia, U.; Kamuni, V.; Sheikh, A.; Wagh, S.; Patel, D. Energy trading of electric vehicles using blockchain and smart contracts. In Proceedings of the 2019 18th European Control Conference, ECC, Naples, Italy, 25–28 June 2019; pp. 3958–3963.
29. Hatim, S.M.; Elias, S.J.; Ali, R.M.; Jasmis, J.; Aziz, A.A.; Mansor, S. Blockchain-based Internet of Vehicles (BIoV): An Approach towards Smart Cities Development. In Proceedings of the 2020 5th IEEE International Conference on Recent Advances and Innovations in Engineering, ICRAIE 2020-Proceeding, Jaipur, India, 1–3 December 2020.
30. Lasla, N.; Al-Ammari, M.; Abdallah, M.; Younis, M. Blockchain Based Trading Platform for Electric Vehicle Charging in Smart Cities. *IEEE Open J. Intell. Transp. Syst.* **2020**, *1*, 80–92. [CrossRef]
31. Ayobi, S.; Wang, Y.; Rabbani, M.; Dorri, A.; Jelodar, H.; Huang, H.; Yarmohammadi, S. A Lightweight Blockchain-Based Trust Model for Smart Vehicles in VANETs. In *Lecture Notes in Computer Science (Including Subseries Lecture Notes in Artificial Intelligence and Lecture Notes in Bioinformatics)*; Wang, G., Chen, B., Li, W., Di Pietro, R., Yan, X., Han, H., Eds.; Springer International Publishing: Berlin/Heidelberg, Germany, 2021; 12382 LNCS; pp. 276–289. ISBN 9783030688509.
32. Tripathi, G.; Ahad, M.A.; Sathiyanarayanan, M. The Role of Blockchain in Internet of Vehicles (IoV): Issues, Challenges and Opportunities. In Proceedings of the 4th International Conference on Contemporary Computing and Informatics, IC3I 2019, Singapore, 12–14 December 2019; IEEE: Piscataway, NJ, USA, 2019; pp. 26–31.
33. Mollah, M.B.; Zhao, J.; Niyato, D.; Guan, Y.L.; Yuen, C.; Sun, S.; Lam, K.Y.; Koh, L.H. Blockchain for the Internet of Vehicles towards Intelligent Transportation Systems: A Survey. *IEEE Internet Things J.* **2021**, *8*, 4157–4185. [CrossRef]
34. Miglani, A.; Kumar, N.; Chamola, V.; Zeadally, S. Blockchain for Internet of Energy management: Review, solutions, and challenges. *Comput. Commun.* **2020**, *151*, 395–418. [CrossRef]
35. Brereton, P.; Kitchenham, B.A.; Budgen, D.; Turner, M.; Khalil, M. Lessons from applying the systematic literature review process within the software engineering domain. *J. Syst. Softw.* **2007**, *80*, 571–583. [CrossRef]
36. Kitchenham, B.A.; Brereton, P.; Turner, M.; Niazi, M.K.; Linkman, S.; Pretorius, R.; Budgen, D. Refining the systematic literature review process-two participant-observer case studies. *Empir. Softw. Eng.* **2010**, *15*, 618–653. [CrossRef]
37. Moher, D.; Liberati, A.; Tetzlaff, J.; Altman, D.G. Preferred reporting items for systematic reviews and meta-analyses: The PRISMA statement. *J. Clin. Epidemiol.* **2009**, *62*, 1006–1012. [CrossRef] [PubMed]
38. Knirsch, F.; Unterweger, A.; Engel, D. Privacy-preserving blockchain-based electric vehicle charging with dynamic tariff decisions. *Comput. Sci.-Res. Dev.* **2018**, *33*, 71–79. [CrossRef]
39. Aggarwal, S.; Kumar, N. A Consortium Blockchain-Based Energy Trading for Demand Response Management in Vehicle-to-Grid. *IEEE Trans. Veh. Technol.* **2021**, *70*, 9480–9494. [CrossRef]
40. Tsao, Y.C.; Van Thanh, V.; Wu, Q. Sustainable microgrid design considering blockchain technology for real-time price-based demand response programs. *Int. J. Electr. Power Energy Syst.* **2021**, *125*, 106418. [CrossRef]
41. Zhang, T.; Pota, H.; Chu, C.C.; Gadh, R. Real-time renewable energy incentive system for electric vehicles using prioritization and cryptocurrency. *Appl. Energy* **2018**, *226*, 582–594. [CrossRef]
42. Lazaroiu, C.; Roscia, M. New approach for smart community grid through blockchain and smart charging infrastructure of evs. In Proceedings of the 8th International Conference on Renewable Energy Research and Applications, ICRERA, Brasov, Romania, 3–6 November 2019; pp. 337–341.
43. Jindal, A.; Aujla, G.S.; Kumar, N.; Villari, M. GUARDIAN: Blockchain-Based Secure Demand Response Management in Smart Grid System. *IEEE Trans. Serv. Comput.* **2020**, *13*, 613–624. [CrossRef]
44. Jiang, Y.; Zhou, K.; Lu, X.; Yang, S. Electricity trading pricing among prosumers with game theory-based model in energy blockchain environment. *Appl. Energy* **2020**, *271*, 115239. [CrossRef]
45. Prabadevi, B.; Pham, Q.-V.; Liyanage, M.; Deepa, N.; VVSS, M.; Reddy, S.; Maddikunta, P.K.R.; Khare, N.; Gadekallu, T.R.; Hwang, W.-J. Deep Learning for Intelligent Demand Response and Smart Grids: A Comprehensive Survey. *arXiv* **2021**, arXiv:2101.08013.
46. Danish, S.M.; Zhang, K.; Jacobsen, H.A.; Ashraf, N.; Qureshi, H.K. BlockEV: Efficient and Secure Charging Station Selection for Electric Vehicles. *IEEE Trans. Intell. Transp. Syst.* **2021**, *22*, 4194–4211. [CrossRef]
47. Guo, Z.; Ji, Z.; Wang, Q. Blockchain-enabled demand response scheme with individualized incentive pricing mode. *Energies* **2020**, *13*, 5213. [CrossRef]
48. Karandikar, N.; Abhishek, R.; Saurabh, N.; Zhao, Z.; Lercher, A.; Marina, N.; Prodan, R.; Rong, C.; Chakravorty, A. Blockchain-based prosumer incentivization for peak mitigation through temporal aggregation and contextual clustering. *Blockchain Res. Appl.* **2021**, *2*, 100016. [CrossRef]
49. Wen, S.; Xiong, W.; Tan, J.; Chen, S.; Li, Q. Blockchain enhanced price incentive demand response for building user energy network in sustainable society. *Sustain. Cities Soc.* **2021**, *68*, 102748. [CrossRef]
50. Karandikar, N.; Chakravorty, A.; Rong, C. Blockchain based transaction system with fungible and non-fungible tokens for a community-based energy infrastructure. *Sensors* **2021**, *21*, 3822. [CrossRef]
51. Pop, C.; Cioara, T.; Antal, M.; Anghel, I.; Salomie, I.; Bertoncini, M. Blockchain based decentralized management of demand response programs in smart energy grids. *Sensors* **2018**, *18*, 162. [CrossRef]
52. Zahid, M.; Ali, I.; Khan, R.J.U.H.; Noshad, Z.; Javaid, A.; Javaid, N. Blockchain Based Balancing of Electricity Demand and Supply. In *Lecture Notes in Networks and Systems*; Barolli, L., Hellinckx, P., Enokido, T., Eds.; Springer International Publishing: Cham, Switzerland, 2020; Volume 97, pp. 185–198. ISBN 978-3-030-33505-2/978-3-030-33506-9.
53. Pop, C.D.; Antal, M.; Cioara, T.; Anghel, I.; Salomie, I. Blockchain and demand response: Zero-knowledge proofs for energy transactions privacy. *Sensors* **2020**, *20*, 5678. [CrossRef]

54. Zhou, Z.; Wang, B.; Guo, Y.; Zhang, Y. Blockchain and Computational Intelligence Inspired Incentive-Compatible Demand Response in Internet of Electric Vehicles. *IEEE Trans. Emerg. Top. Comput. Intell.* **2019**, *3*, 205–216. [CrossRef]
55. Al-Obaidi, A.; Khani, H.; Farag, H.E.Z.; Mohamed, M. Bidirectional smart charging of electric vehicles considering user preferences, peer to peer energy trade, and provision of grid ancillary services. *Int. J. Electr. Power Energy Syst.* **2021**, *124*, 106353. [CrossRef]
56. Kumari, A.; Tanwar, S. A Data Analytics Scheme for Security-aware Demand Response Management in Smart Grid System. In Proceedings of the 2020 IEEE 7th Uttar Pradesh Section International Conference on Electrical, Electronics and Computer Engineering (UPCON), Prayagraj, India, 27–29 November 2020.
57. Borges, C.E.; Kapassa, E.; Touloupou, M.; Macón, J.L.; Casado-Mansilla, D. Blockchain application in P2P energy markets: Social and legal aspects. *Connect. Sci.* **2022**, *34*, 1066–1088. [CrossRef]
58. Mengelkamp, E.; Gärttner, J.; Rock, K.; Kessler, S.; Orsini, L.; Weinhardt, C. Designing microgrid energy markets: A case study: The Brooklyn Microgrid. *Appl. Energy* **2018**, *210*, 870–880. [CrossRef]
59. Duan, Q.; Quynh, N.V.; Abdullah, H.M.; Almalaq, A.; Duc Do, T.; Abdelkader, S.M.; Mohamed, M.A. Optimal Scheduling and Management of a Smart City within the Safe Framework. *IEEE Access* **2020**, *8*, 161847–161861. [CrossRef]
60. Korkmaz, A.; Kılıç, E.; Türkay, M.; Çakmak, Ö.F.; Arslan, T.Y. A Blockchain Based P2P Energy Trading Solution for Smart Grids. Researchgate.net. 2021. Available online: https://www.researchgate.net/profile/Ulas-Erdogan/publication/349961429_A_Blockchain_Based_P2P_Energy_Trading_Solution_for_Smart_Grids/links/604942ff299bf1f5d83d8b5d/A-Blockchain-Based-P2P-Energy-Trading-Solution-for-Smart-Grids.pdf (accessed on 10 March 2022).
61. Durillon, B.; Davigny, A.; Kazmierczak, S.; Barry, H.; Saudemont, C.; Robyns, B. Decentralized neighbourhood energy management considering residential profiles and welfare for grid load smoothing. *Sustain. Cities Soc.* **2020**, *63*, 102464. [CrossRef]
62. Tushar, W.; Saha, T.K.; Yuen, C.; Smith, D.; Poor, H.V. Peer-to-Peer Trading in Electricity Networks: An Overview. *IEEE Trans. Smart Grid* **2020**, *11*, 3185–3200. [CrossRef]
63. Long, C.; Wu, J.; Zhou, Y.; Jenkins, N. Peer-to-peer energy sharing through a two-stage aggregated battery control in a community Microgrid. *Appl. Energy* **2018**, *226*, 261–276. [CrossRef]
64. Inayat, K.; Hwang, S.O. Load balancing in decentralized smart grid trade system using blockchain. *J. Intell. Fuzzy Syst.* **2018**, *35*, 5901–5911. [CrossRef]
65. Pinto, R.; Bessa, R.J.; Sumaili, J.; Matos, M.A. Distributed multi-period three-phase optimal power flow using temporal neighbors. *Electr. Power Syst. Res.* **2020**, *182*, 106228. [CrossRef]
66. Chen, S.; Liu, C.C. From demand response to transactive energy: State of the art. *J. Mod. Power Syst. Clean Energy* **2017**, *5*, 10–19. [CrossRef]
67. Abidin, A.; Aly, A.; Cleemput, S.; Mustafa, M.A. Secure and Privacy-Friendly Local Electricity Trading and Billing in Smart Grid. *arXiv* **2018**, arXiv:1801.08354.
68. Lei, L.; Taorong, G.; Jindou, Y.; Feixiang, G.; Tao, X.; Tao, C.; Songsong, C. Research on the strategy of adjustable load resources participating in distributed trading market. In Proceedings of the 2021 IEEE 2nd International Conference on Big Data, Artificial Intelligence and Internet of Things Engineering, ICBAIE, Nanchang, China, 26–28 March 2021; pp. 788–792.
69. Cruz, C.; Palomar, E.; Bravo, I.; Gardel, A. Cooperative demand response framework for a smart community targeting renewables: Testbed implementation and performance evaluation. *Energies* **2020**, *13*, 2910. [CrossRef]
70. Veras, J.M.; Silva, I.R.S.; Pinheiro, P.R.; Rabêlo, R.A.L. Towards the handling demand response optimization model for home appliances. *Sustainability* **2018**, *10*, 616. [CrossRef]
71. Kermani, M.; Parise, G.; Shirdare, E.; Martirano, L. Transactive Energy Solution in a Port's Microgrid based on Blockchain Technology. In Proceedings of the 2020 IEEE International Conference on Environment and Electrical Engineering and 2020 IEEE Industrial and Commercial Power Systems Europe, EEEIC / I and CPS Europe, Madrid, Spain, 9–12 June 2020; pp. 1–6.
72. Golpîra, H.; Bahramara, S. Internet-of-things-based optimal smart city energy management considering shiftable loads and energy storage. *J. Clean. Prod.* **2020**, *264*, 121620. [CrossRef]
73. Samuel, O.; Javaid, N.; Shehzad, F.; Iftikhar, M.S.; Iftikhar, M.Z.; Farooq, H.; Ramzan, M. Electric Vehicles Privacy Preserving Using Blockchain in Smart Community. In *Lecture Notes in Networks and Systems*; Barolli, L., Hellinckx, P., Enokido, T., Eds.; Springer International Publishing: Cham, Switzerland, 2020; Volume 97, pp. 67–80. ISBN 978-3-030-33505-2/978-3-030-33506-9.
74. Ramos, D.; Khorram, M.; Faria, P.; Vale, Z. Load Forecasting in an Office Building with Different Data Structure and Learning Parameters. *Forecasting* **2021**, *3*, 242–255. [CrossRef]
75. Gilleran, M.; Bonnema, E.; Woods, J.; Mishra, P.; Doebber, I.; Hunter, C.; Mitchell, M.; Mann, M. Impact of electric vehicle charging on the power demand of retail buildings. *Adv. Appl. Energy* **2021**, *4*, 100062. [CrossRef]
76. Guo, N.; Zhang, X.; Zou, Y.; Guo, L.; Du, G. Real-time predictive energy management of plug-in hybrid electric vehicles for coordination of fuel economy and battery degradation. *Energy* **2021**, *214*, 119070. [CrossRef]
77. Wu, Y.; Wu, Y.; Guerrero, J.M.; Vasquez, J.C. Decentralized transactive energy community in edge grid with positive buildings and interactive electric vehicles. *Int. J. Electr. Power Energy Syst.* **2022**, *135*, 107510. [CrossRef]

Article

SASLedger: A Secured, Accelerated Scalable Storage Solution for Distributed Ledger Systems

Haoli Sun [1,*], Bingfeng Pi [1], Jun Sun [2], Takeshi Miyamae [3] and Masanobu Morinaga [3]

1. Fujitsu R&D Center Co., Ltd., Suzhou 215123, China; winter.pi@fujitsu.com
2. Fujitsu R&D Center Co., Ltd., Beijing 100022, China; sunjun@fujitsu.com
3. Fujitsu Limited, Kawasaki 211-8588, Japan; miyamae.takeshi@fujitsu.com (T.M.); morinaga@fujitsu.com (M.M.)
* Correspondence: sunhaoli@fujitsu.com; Tel.: +86-512-62925255

Abstract: Blockchain technology provides a "tamper-proof distributed ledger" for its users. Typically, to ensure the integrity and immutability of the transaction data, each node in a blockchain network retains a full copy of the ledger; however, this characteristic imposes an increasing storage burden upon each node with the accumulation of data. In this paper, an off-chain solution is introduced to relieve the storage burden of blockchain nodes while ensuring the integrity of the off-chain data. In our solution, an off-chain remote DB server stores the fully replicated data while the nodes only store the commitments of the data to verify whether the off-chain data are tampered with. To minimize the influence on performance, the nodes will store data locally at first and transfer it to the remote DB server when otherwise idle. Our solution also supports accessing all historical data for newly joined nodes through a snapshot mechanism. The solution is implemented based on the Hyperledger Fabric (HLF). Experiments show that our solution reduces the block data for blockchain nodes by 93.3% compared to the original HLF and that our advanced solution enhances the TPS by 9.6% compared to our primary solution.

Keywords: blockchain; scalability; storage; data integrity; performance

1. Introduction

Blockchain systems are categorized into permissionless blockchains (a.k.a. a public blockchain) and permissioned blockchains (a.k.a. a consortium/private blockchain). In a permissionless blockchain system, any computer or user that can access the blockchain network can join it or quit at will. The nodes and users of the permissionless blockchain networks are identified by their public keys. Cryptocurrencies such as Bitcoin [1], Ethereum [2], and EOS [3] are constructed as permissionless blockchain systems. Permissioned blockchains are always used for information-sharing among several stakeholders. The nodes and users that want to join the network need to be authorized. Certification Authority (CA) can be used to perform the authorization. HLF [4] and Quorum [5] are typical permissioned blockchain systems.

Blockchain provides a tamper-proof distributed ledger by mechanisms such as decentralized architecture, consensus algorithm, asymmetric encryption, and so on. Transactions that are sent to the blockchain system are packed into blocks by certain rules. A block contains a block header and a block body, the transactions are recorded in the block body. Each block contains the hash of its previous block, so that the blocks form a chain structure, which means that tampering with a historical transaction can be obtained by comparing the block hash saved in the next block. The blocks are usually stored as files in the file system of blockchain nodes. Typically, each node of the blockchain network retains a full copy of the entire chain of blocks, so that each node can check the integrity of the data locally.

Since the size of the chain of blocks increases continually, the blockchain systems are facing a storage scalability issue. For permissionless blockchains, the data size of a

Bitcoin node has reached 366.51 GB [6] and the data size of a Ethereum node has reached 987.54 GB [7] by 28 September 2021; for permissioned blockchains, the issue has also caused concern [8,9]. The storage burden is becoming more onerous for individual participants who run blockchain nodes with their personal computers. This situation may injure the decentralized character of blockchain systems since only wealthy individuals or organizations can afford the increasing storage scaling demand.

Methods proposed to solve the storage scalability issue of blockchain systems are divided into "on-chain" solutions and "off-chain" solutions. On-chain solutions reduce the contents stored in each node by altering the range of consensus or the contents of transactions [10–13]. Off-chain solutions provide off-chain storage devices to store data, but most of them focus on reducing the data size before the data are uploaded to blockchains [14–17]. This approach uses blockchain as a proof repository to ensure the integrity of off-chain data, but the off-chain data cannot be retrieved through blockchain node, on the contrary, block archiver [18] reduces the data size of the data that has been uploaded to the blockchain and stored in the blockchain nodes, but block archiver fails to ensure the integrity of the data stored off-chain, which severely damages the tamper-proof property of the original blockchain system. To bridge this gap, an off-chain solution for the data that has been uploaded to the blockchain with tamper-proof property is proposed in this paper. The data are eliminated from the blockchain node and transmitted to a remote DB server while the node saves a concise vector commitment (VC) [19] to ensure the integrity of the data stored off-chain.

The contributions made by the present research are as follows:

1. An off-chain solution to solve the data scalability issue of blockchain systems while ensuring the integrity of the data stored off-chain is proposed, the target of this solution is to reduce the size of the data that has been uploaded to the blockchain and stored in the blockchain nodes. Each node saves a concise VC to ensure the integrity of the data of a block while the raw data are eliminated from the node.
2. The performance of our solution is improved compared to our primary solution "xFabLedger" [20] by separating the transaction-processing phase and the data-reduction phase. The data reduction is performed when a blockchain node is backed up to avoid read-write conflicts and negative impact on performance.
3. The solution is implemented based on Hyperledger Fabric (HLF) V2.3.2 [21], and storage experiments and performance experiments are conducted to prove the effectiveness of our solution.

The rest of the paper is organized as follows: Section 2 describes different approaches used when solving the scalability issues of blockchain. Section 3 introduces the transaction-processing model and storage structure of the HLF blockchain system and the mechanism of VC. Section 4 introduces our solution and then analyzes the security and performance features thereof. Section 5 demonstrates the details of our implementation based on HLF. Section 6 shows our experimental results and their evaluations. Section 7 concludes with recommendations for future research.

2. Related Work

Blockchain technology is not only underpinning cryptocurrencies, but also widely adopted in other industries, such as finance [22–24], IoT [14,25,26] and healthcare [27–29]. Although blockchain benefits the systems that use this technology, it brings scalability issues to them [30–32]. The methods to solve the storage scalability issue of blockchain are categorized into "on-chain" solutions and "off-chain" solutions.

2.1. On-Chain Solutions

"Sharding" is a method that divides the nodes of the blockchain network into subgroups called shards, each shard acts as an independent blockchain network to process transactions and store data [10,11]. This approach reduces the data need to store for blockchain nodes, but each node still needs to store the full copy of the blocks of the subgroup.

Several works propose storing the data distributedly similar to traditional peer-to-peer content distribution networks [33–35] do. CUB [12] introduces the concept of "Consensus Unit" (CU), in which nodes combine their resources to maintain the blockchain data together rather than based on one copy per node. The differences between a CU and a shard are: (1) as a whole, each CU stores identical data while each shard stores their unique data; (2) each node in a shard stores identical data while each node in CU stores different blocks. Jidar [13] proposes a method allowing Bitcoin blockchain nodes to store only those transactions that are related to themselves. To verify a transaction, a proof is attached with the transaction when it is sent. Moreover, a bloom filter is added to each block to check whether a transaction has been affected.

On-chain solutions can reduce the storage burden of blockchain, but each node still needs to store at least part of the blocks. On-chain solutions also tend to introduce extra network overhead since the nodes must communicate to decide which node stores which part of the entire data.

2.2. Off-Chain Solutions

Most of the off-chain solutions focus on reducing the data size before the data are uploaded to blockchains [14–17]. In these solutions, blockchains work as proof repositories to ensure the integrity of off-chain data, but the problems are: (1) off-chain data cannot be retrieved through a blockchain node; (2) each node still need to store the full copy of the blocks.

Block archiver [18] is a solution used to reduce the data size of the data that has been uploaded to the blockchain and stored in the blockchain nodes by transmitting the data to an off-chain block archiver repository. It addresses the storage issue for HLF. In an organization (formed by serveral blockchain nodes) of HLF, a block archiver repository is deployed off-chain to store archived block files, a "block archiver" is deployed on the anchor/leader peer (a kind of blockchain node), and "block archiver clients" are deployed on other peers. The "block archiver" is responsible for transmitting block files from the peer's local file system to the block archiver repository, deleting local block files and notifying the "block archiver clients" to delete relevant block files from their local file systems. Fast fabric [36] also mentioned an approximate idea that storing the blocks in a distributed storage cluster. However, these solutions have not considered the method to keep the integrity of the block data after they are stored off-chain.

Our solution is proposed to address the aforementioned problems in current solutions. A remote DB server is used to store the data that has been stored in the blockchain nodes, while the off-chain stored data are still retrievable through blockchain nodes. Concise commitments are stored in each node to ensure the data integrity for the blocks, therefore the size of data that stored on blockchain nodes are extremely reduced. The concept of CU is also adopted to avoid centralization and reduce network overhead.

3. Background

To help understand our solution better, several relevant technologies are introduced as background in this section.

3.1. Hyperledger Fabric
3.1.1. A Modular Pluggable Blockchain System

Hyperledger Fabric is a popular open-sourced permissioned blockchain system. In a permissionless blockchain system, the fundamental components (e.g., consensus algorithm and block generation rules) must be pre-determined before the system starts to provide services, since each node in a blockchain system must obey the same rules. If a change is needed after the blockchain system begins to run, the so-called "fork" [37] operation is required. On the contrary, components in HLF are modular and pluggable. Provided the stakeholders reach a consensus, changes to the system can be easily realized. The nodes in HLF networks are divided by their different functions. The nodes that are responsible for saving blocks and answering query requests are called peers, some of the peers are also obtained to carry out the calculations defined in the transactions, these peers are called endorsers. The peers in the network always belong to different organizations: each organization represents one or several stakeholders in the real world. The nodes that are responsible for ordering the transactions and packing them into blocks are called orderers, it is recommended that each organization deploys one orderer to represent the organization. The orderers provide ordering services together follow certain consensus algorithms such as Raft [38] or PBFT [39]. The orderers are referred to as Ordering Service Nodes (OSN). Figure 1 illustrates the structure of a typical HLF network.

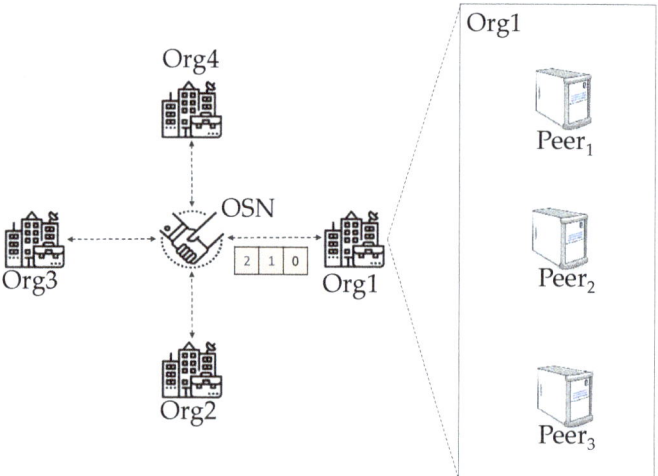

Figure 1. The structure of Hyperledger Fabric network. The Ordering Service Nodes (OSN) order blocks and deliver ordered blocks to each organization (Org). Each organization contains several peers.

3.1.2. Transaction-Processing Architecture

For most of the blockchain systems, a consensus of the order of transactions in a block must be reached before calculating the results of the transactions. Since each node will calculate the results independently to verify the results, the order of the transactions must be determined to maintain consistency among nodes. This architecture renders the computational resources of the blockchain system un-scalable. To solve this problem, HLF adopts a different transaction-processing architecture called "execute-order-validate" [4], in which transactions are calculated concurrently by multiple endorsers then ordered by OSN and later validated by peers.

3.1.3. Storage Structure of A Peer

The storage structure of an HLF peer comprises a chain of blocks and several local databases. In HLF, a block consists of three parts [21]:
- *Block Header* comprises three fields
  - *Block Number* is the sequence of the current block. The block number is counted from 0. The first block is called genesis block.
  - *Data Hash* is the hash value of all the transactions that are recorded in the current block.
  - *Previous Hash* is the Data Hash of its previous block.

- Block Data contains the ordered transactions. The transaction creators' and the endorsers' signatures are attached with each transaction. The count of transactions stored in a block is called the "block size" which can be configured before the network is started.
- Block Metadata contains the validation codes used to verify transactions and the signature of the block creator.

The blocks are chained together through the "Previous hash", and they are stored in the file system as block files. The size of block files is configured before the network is started so that the number of blocks that each block file contains is determined.

The blocks preserve all original information pertaining to each transaction, but for data retrieval, local databases are needed to provide the necessary functionality:

- StateDB records the newest status of all the objects defined in the applications of the blockchain system. The status is recorded as <Key, Value> pairs. It can be chosen that whether CouchDB [40] or LevelDB [41] is used as the state DB before the HLF network is started.
- IndexDB indicates where to find each block and transaction in the block files. It records the sequence number of the block file retaining the indexed blocks and transactions and offsets of those blocks and transactions.
- HistoryDB can be used to track each change for each key. The sequence numbers of key-related blocks and transactions are recorded.

3.1.4. Snapshot

A snapshot mechanism [42] is officially included in the HLF V2.3 to back-up a peer. A snapshot comprises several files that are exported from local databases of a peer. It can be used for checking whether ledger forks occur among different organizations and letting a new peer join the channel without synchronizing previously committed blocks from other peers. When a peer is going to generate a snapshot, the committing of blocks is stopped to prevent read-write conflicts.

3.2. Vector Commitment

For a given element e and a sequence number i, VC [19] can be used to identify whether e is equal to the ith element of an ordered set $\mathbf{e}$ $(e_1, \ldots e_n)$. In blockchain systems, the transactions in a block form an ordered set of elements, thus VC can be used to identify whether a given transaction is in a block and its position is also correctly provided. VC demonstrates the features of hiding, position binding, conciseness and an ability to be updated. The following six algorithms are used to define VC:

1. Given security parameter k and the size n of ordered set $\mathbf{e}$, the "KeyGen" algorithm outputs the public parameters pp.

$$pp \leftarrow KeyGen(1^k, n) \quad (1)$$

2. Given the public parameters pp and the ordered set $\mathbf{e}$, the "Commitment" algorithm outputs the commitment $\mathbf{C}$ of $\mathbf{e}$.

$$\mathbf{C} \leftarrow Commitment_{pp}(e_1, \ldots, e_n) \quad (2)$$

3. The "Open" algorithm is used to generate the proof π_i for the ith element.

$$\pi_i \leftarrow Open_{pp}(i, e_1, \ldots, e_n) \quad (3)$$

4. The "Verify" algorithm is employed to check whether the given element e is equal to e_i through the proof π_i. If the proof is accepted, the algorithm outputs 1.

$$\{0, 1\} \leftarrow Verify_{pp}(\mathbf{C}, i, e, \pi_i) \quad (4)$$

5. When the ith element of **e** needs to be updated with e', the "Update" algorithm can be run to output a new commitment C' and the update information U.

$$C', U \leftarrow Update_{pp}(C, e, e', i) \qquad (5)$$

6. When the ith element of **e** is updated, the proofs of other elements (e.g., e_j) also need to be updated. The algorithm "ProofUpdate" outputs the updated proof π'_j of element e_j.

$$\pi'_j \leftarrow ProofUpdate_{pp}(\pi_j, e', i, U) \qquad (6)$$

4. SASLedger: A Secured, Accelerated Scalable Storage Solution

To relieve the storage burden for blockchain nodes, we focus on reducing the size of the chain of blocks since it alone is responsible for most of the storage burden of a blockchain node. Figure 2 shows the system architecture of our solution. An off-chain solution within each CU is used to achieve lower network overhead compared to on-chain solutions or solutions without adopting CU [13] or solutions without adopting CU [10–13]: an off-chain remote DB server for each CU is added to store the blocks; VC is used to ensure the integrity of each transaction in a block stored in the remote DB server. Although the blocks are stored in the remote DB server, the nodes still maintain their local databases respectively and still be able to check the data integrity of the blocks that they transmit to the remote DB server, thus the decentralized character and tamper-proof character of blockchain system are retained.

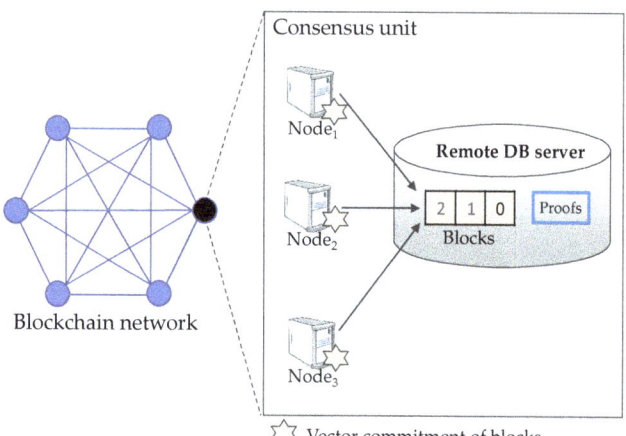

Figure 2. System architecture of our solution. The entire blockchain network is divided into multiple Consensus Unit (CU), in which nodes combine their resources to maintain the blockchain data together.

4.1. Proving Data Integrity through VC

To prove the integrity of the data in a block, Jidar [13] attaches a proof for each transaction, which requires extra storage space and generates significant network overhead. Compared to Jidar, in our solution, nodes only store a concise VC of a block, the proof for a transaction is generated and provided by the remote DB server only when the transaction is requested.

A Merkle tree [43] is used to generate VC for the set of transactions in a block and proofs for the transactions. As shown in Figure 3, the hashes of the transactions in a block $(tx_1, \ldots, tx_n)$ form the leaf nodes of the Merkle tree, the root of the Merkle tree is considered to be VC (**C**) of the block (7). When the proof of the ith transaction tx_i is requested, the list which includes adjacent sibling nodes of the corresponding leaf and its ancestor nodes is returned as the proof π (8), in Figure 3, the list $[Hash_3, Hash_{12}, Hash_{5678}]$ (shown in yellow) is the proof of transaction tx_4.

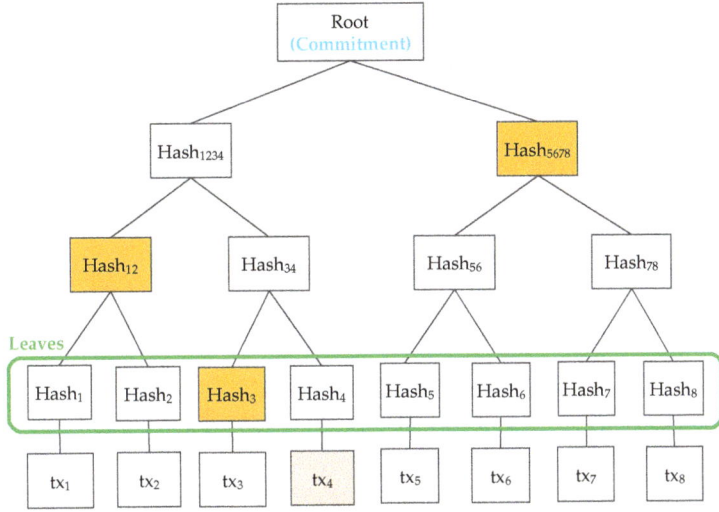

Figure 3. Merkle tree-based vector commitment. The hashes of the transactions in a block $(tx_1, \ldots, tx_n)$ form the leaf nodes of the Merkle tree, the root of the Merkle tree is considered to be VC of the block. The list $[Hash_3, Hash_{12}, Hash_{5678}]$ (shown in yellow) is the proof of transaction tx_4.

When a transaction is requested, the transaction is retrieved from the remote DB server with its proof and verified by calculating the Merkle tree root using the received proof and comparing the root with the previously stored commitment (9). If the root equals the VC, the node can believe that the transaction has not been subject to tampering on the remote DB server. In the case of a block being requested, the node retrieves the block from the remote DB server and validates its integrity by calculating its hash and comparing it with the previously stored hash of the block. The size of a VC is 32 B when SHA256 is used as the hash calculating algorithm while the size of a block with default settings containing 10 transactions is 34 KB [20]. Since the size of a VC is much less than the size of a block, storing its VC instead of the original block can save much storage space.

$$C \leftarrow Commitment(tx_1, \ldots, tx_n) \quad (7)$$

$$\pi \leftarrow Open(i, tx_1, \ldots, tx_n) \quad (8)$$

$$\{0,1\} \leftarrow Verify(C, i, tx, \pi) \quad (9)$$

Since the transactions in a block are fixed, the "Update" and "ProofUpdate" algorithms in conventional VC definition are not applicable. The "KeyGen" algorithm is unnecessary neither since there are no other public parameters to be generated. The algorithms are shown in detail with Algorithms A1–A4 in Appendix A.

4.2. A Primary Solution: xFabLedger

xFabLedger [20] is our primary solution in which after each node receives a block, the node generates commitment for the block and transmits the block to the remote DB server. The block is never stored in the file system of the node. Figure 4 illustrates how a block is processed. xFabLedger solves the excessive storage growth issue and security issue, but experiments indicates that the performance is decreased compared to that of the original blockchain system.

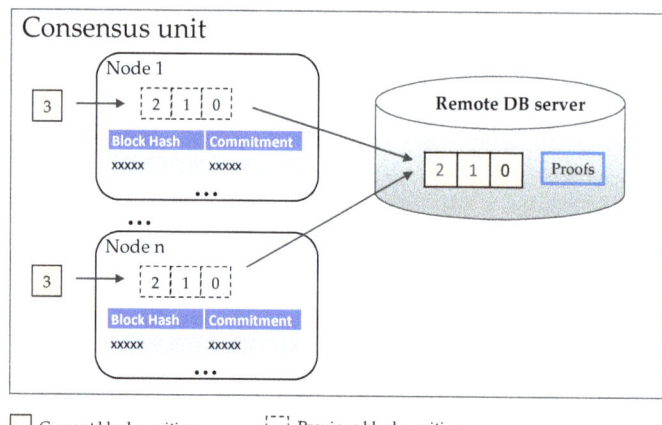

Figure 4. The primary solution. After a block is received by a blockchain node, the node generates commitment for the block and transmits the block to the remote DB server immediately.

4.3. The Advanced Solution: SASLedger

In our advanced solution, the data-reduction operation is separated from the transaction processing phase. The data-reduction operation should be performed when the system is not busy or when the system is being backed up (e.g., the snapshot mechanism in HLF). Figure 5 shows the two phases of our solution: during the transaction-processing phase, the nodes process the transactions as vanilla blockchain nodes do and save the blocks locally, thus the TPS remains unaffected. During the data-reduction phase, the nodes perform the following operations for each committed block:

- Calculate a VC for the set of transactions contained in the block;
- Save the hash value of the block and its commitment to local storage as a $\langle Key, Value \rangle$ pair which is used as an index of the block: $\langle blockHash, commitment \rangle$;
- Transmit the content of the block to the remote DB server;
- Delete the block from its local storage.

When a query is sent to a node to retrieve a block or a transaction, the node retrieves the block or transaction from the remote DB server and transmits it to the client after verification.

A snapshot of the blocks can be generated from the block indices: this can be used to build block indices for a newly joined node in the future.

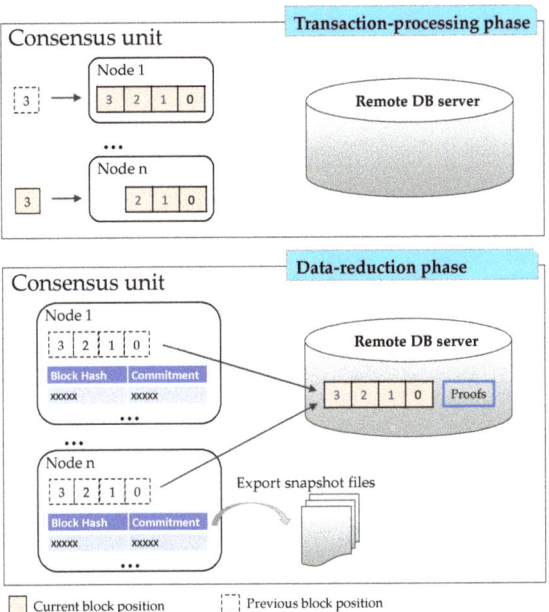

Figure 5. Two phases of our advanced solution. In the transaction-processing phase, the nodes process the transactions as vanilla blockchain nodes do and save the blocks locally. In the data-reduction phase, the nodes generate commitments for the blocks, transmit the blocks to the remote DB server, and delete the blocks locally.

5. Implementation Based on Hyperledger Fabric

Our advanced solution is implemented based on HLF V2.3.2. The reasons why HLF is chosen as the basis for our implementation are:

1. We are focusing on enterprise-level applications while HLF is a popular permissioned blockchain solution for information-sharing among companies due to its high performance and rich privacy preserving mechanisms [4,21];
2. The nodes in a HLF network are already divided into "organizations", which is suitable for building a CU since the peers in an organization are managed by the same administrator, and they are identified by each other through digital signatures;
3. The snapshot mechanism introduced in HLF V2.3.2 provides a perfect timing to perform the "data-reduction" operation of our solution.

A remote DB server is deployed in each organization of HLF to store all the blocks previously stored in the peers in this organization. When a new generated block is arrived, the peers commit the block (saving the block to its file system and saving the relevant information to its local databases) as vanilla HLF peers do, so that the TPS of the blockchain system is unaffected. The data-reduction operation is conducted when a snapshot is generated.

5.1. Peer Node

As mentioned in Section 3, a peer node of HLF performs the following operations:

- *Endorsing*: performing the calculations defined in transactions;
- *Committing*: preserving blocks to the file system of the peer and preserving some block-related information to the peer's local databases;
- *Responding query requests*: searching blocks and local databases to find information that is requested by users and answering the requests;
- *Generating snapshot*: stop committing and then exporting information from local databases to snapshot files.

The "Generating snapshot" operation is modified to transmit the blocks to the remote DB server and delete the blocks from its local file system. The "Responding query requests" operation is also modified by adding verification process after the blocks and transactions are retrieved from the remote DB server. The communications between peers and the remote DB server are realized by way of a GRPC protocol [44]. The "Endorsing" and "Committing" operations remain unmodified.

5.1.1. Modification of Generating Snapshot

Before a block is transmitted to the remote DB server, the commitment of the block is generated and the indexDB of the peer is updated, so that the peer can retrieve transactions and blocks from the remote DB server after transmitting the blocks. A comparison of contents in indexDB before and after the update is provided (Table 1).

Table 1. Contents of indexDB before and after updating.

Key	Before Update	After Update
blkHash	block file, block offset	VC for the block, hash of the entire block [a]
blkNum	block file, block offset	blkHash
txID, blkNum, txNum	block file, block offset, tx offset, tx validation code	tx validation code

[a] MetaData is not included when blockHash is calculated.

In HLF implementation, a peer who joins a channel through a snapshot can obtain the latest world states when the snapshot is generated; but it cannot access the contents of blocks that have been committed before the snapshot is generated since the local databases of the peers that generate the snapshot only contain information of those blocks that are stored locally. However, in our implementation, since the indexDB is updated for the blocks that are going to be transmitted to the remote DB server, those blocks can be accessed by new peers if they join the channel through copying the information from the snapshot to their local databases. To support this function fully, the snapshot contents exported from different local databases are extended: besides the contents exported from stateDB and configHistoryDB which are also included in the snapshot of HLF, the snapshot of SASLedger also contains the history records of states that are exported from the historyDB and all the <Key, Value> records in the indexDB after it is updated.

5.1.2. Modification of "Responding Query Requests"

In HLF, peers provide four query interfaces for users:

- *QueryTransaction* is used to retrieve a transaction-by-transaction ID;
- *QueryBlock* is employed to retrieve a block-by-block number;
- *QueryBlockByHash* is to obtain a block-by-block hash;
- *QueryBlockByID* is to acquire a block using the transaction ID of one of the transactions that is compacted within the block.

Since SASLedger retrieves blocks and transactions from the remote DB server, the retrieved data must be verified to ensure that they are not subject to tampering on the remote DB server. The verification is performed by the peer as described in Section 4.1.

5.2. Remote DB Server

The remote DB server is implemented based on leveldb. It comprises three databases: blockDB, txIndexDB, and merkleTreeDB. When a block is received by the remote DB server, first, the redundancy of the block is checked through its hash; secondly, if the received block has not already been saved, the blockData are saved to the blockDB, the transactions in the block body are traversed to save the indices of the transactions to the txIndexDB, and the Merkle Tree is calculated for the block and saved to the merkleTreeDB. Table 2 lists the contents of these databases.

Table 2. Contents of remote databases.

Database	Key	Value
blockDB	blkHash	blockData
txIndexDB	txID	blkHash, txNum [a]
merkleTreeDB	merkleNode	adjacent sibling node

[a] The sequence number that the tx is stored in the block.

The remote DB server provides four APIs for the peers to invoke:
- *AddBlock*: uploads a block to remote DB server;
- *GetBlockByHash*: retrieves a block by blockHash and returns the block;
- *GetBlockByTxId*: retrieves a block by transaction ID and returns the block;
- *GetTransactionByTxId*: retrieves a transaction-by-transaction ID and returns the transaction and its proof.

6. Experiment and Evaluation

Experiments are conducted to elucidate how much storage space the SASLedger can save compared with HLF. Meanwhile, experiments are performed to compare the TPS and query latencies of SASLedger, xFabLedger, and HLF.

6.1. Basic Experimental Settings

Table 3 shows the configuration of the testing machines that we use to perform these experiments. We use six computers, each with identical configurations: three of them are used as blockchain nodes, one of them as the remote DB server. and the other two as Performance Traffic Engine (PTE) [45] machines for sending transactions to the blockchain networks. Table 4 lists the settings for blockchain networks.

Table 3. Testing machine configuration.

Configuration Item	Value
OS	Ubuntu 16.04
CPU	i5-9400 (2.9 GHz) 6 cores
Memory	8 GB
Hard Disk	SSD 512 GB
Network Bandwidth	1000 Mbps

Table 4. Blockchain network settings.

Blockchain Network Component	Value
Orderer node number	One
Consensus algorithm	Raft
Organization number	Two
Peer number in each organization	One
StateDB	leveldb

6.2. Storage Experiments

Experiments are carried out to understand to what level the SASLedger can relieve the storage burden of a blockchain node. In Figure 6, the grey bars and red bars show the block file storage demand of HLF and SASLedger, respectively; the yellow bars indicate the disk size of the remote DB server; the green line shows the storage consumption reduction rate (10) of SASLedger compared with HLF. When the number of transactions reaches 480,000, the reduction rate of peer block file size reaches 93.3%. Since the first block file containing the genesis block and the latest block file are kept when deleting block files in peers, the SASLedger holds storage consumption records (red bars in Figure 6) of a constant value when the number of transactions changes.

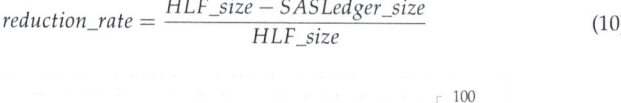

$$reduction\_rate = \frac{HLF\_size - SASLedger\_size}{HLF\_size} \qquad (10)$$

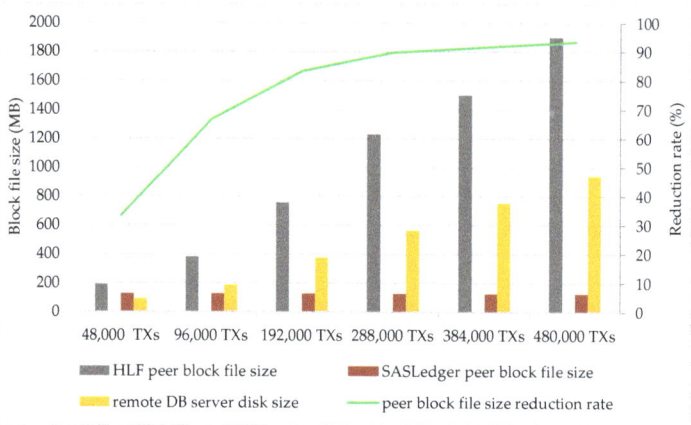

Figure 6. Storage comparison between HLF and SASLedger. The grey bars and red bars show the block file storage demand of HLF and SASLedger, respectively; the yellow bars indicate the disk size of the remote DB server; the green line shows the storage consumption reduction rate of SASLedger compared with HLF.

6.3. Performance Experiments

To assess the performance of blockchain networks, two PTE machines are used to send a certain number of write-only transactions to the two organizations. The configuration of the PTE is summarized in Table 5.

Table 5. PTE settings.

Setting Item	Value
PTE machine number	Two
Process number per PTE	Eight
Tx number per Process	3000

The TPS values of HLF, xFabLedger, and SASLedger are compared when they adopt different block sizes. Figure 7 shows the experimental results. For each of the three blockchain systems, the TPS increases as the block size increases. The reason for this is that over the same number of transactions, the larger the block size, the fewer blocks are generated thus the number of block-oriented operations is reduced. Since xFabLedger transmits the block to the remote DB server during the transaction-processing phase, its TPS is lower than that when using HLF. SASLedger separates the block transmission from transaction-processing phase, which leads to a TPS improvement of 9.6% compared with xFabLedger.

Figure 8 shows the block commitment latency comparison of HLF, xFabLedger, and SASLedger when the block size is changed. The block commitment latency is the time at which a block and its relevant information are recorded. It can be divided into three parts:

- *BlockCommitTime* is the time span to record the block itself;
- *StateCommitTime* is the time span to write relevant information to stateDB;
- *HistoryCommitTime* is the time span to write relevant information to historyDB.

Figure 8 indicates that the total block commitment latency of xFabLedger with each block size is the highest among the three blockchain systems and the commitment latencies of HLF and SASLedger are similar. This observation can explain the result shown in Figure 7.

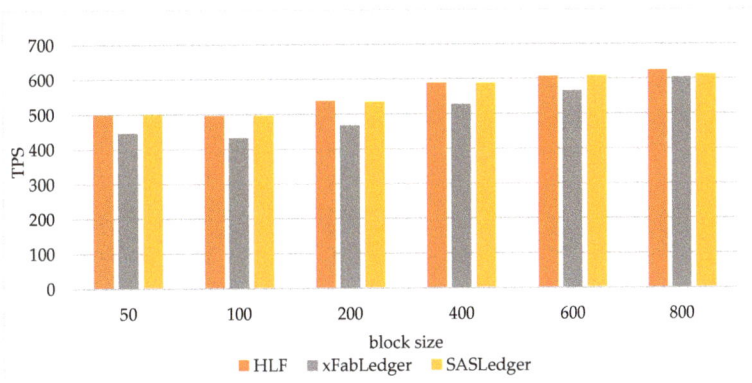

Figure 7. Throughput comparison of HLF, xFabLedger, and SASLedger. The TPS of xFabLedger is lower than that when using HLF. The TPS is improved to the same level with HLF in SASLedger.

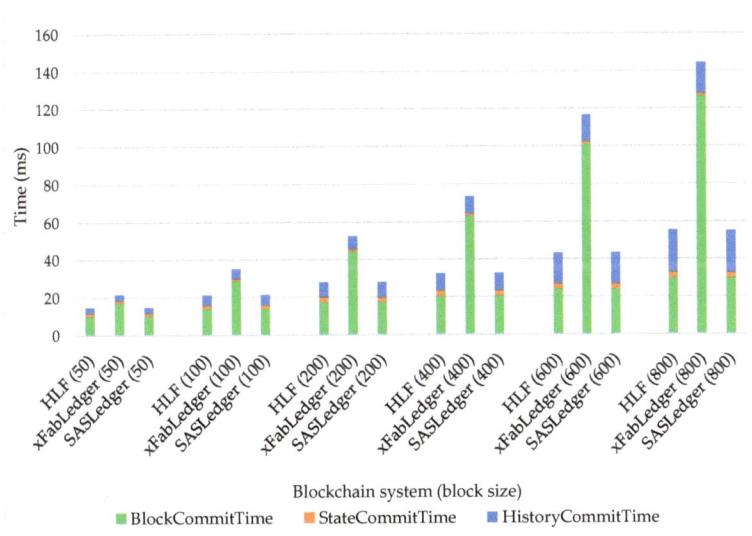

Figure 8. Block commitment latency comparison: HLF, xFabLedger, and SASLedger. The total block commitment latency of xFabLedger with each block size is the highest among the three blockchain systems and the commitment latencies of HLF and SASLedger are similar.

For the performance of query requests, the query latencies of the three blockchain systems are tested by calling the four query interfaces: QueryTransaction, QueryBlock, QueryBlockByID, and QueryBlockByID.

In SASLedger a block is stored in the peers before it is transmitted to the remote DB server. Figure 9 shows that if a block is still stored in the local file system of a peer, the time to query it and the transactions therein is similar to the time cost when using HLF, while for a block that has been transmitted to the remote DB server, the query latency is similar to that when using xFabLedger. Since the blocks are either stored locally in the peers or stored in the remote DB server, the overall query performance is significantly affected by the distribution of stored locations. In the worst situation that all blocks are stored in the remote DB server, the block-query-latency and transaction-query-latency of SASLedger are 31% and 67% higher than those of HLF, respectively.

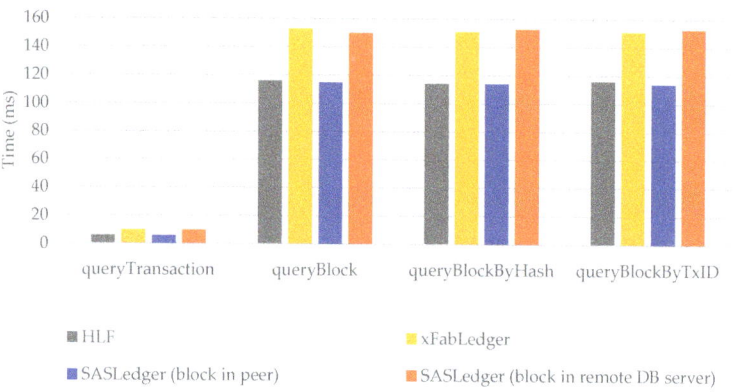

Figure 9. Query latency comparison: HLF, xFabLedger, and SASLedger. When using SASLedger, for a block stored in the local file system of a peer, the time to query it and the transactions therein is similar to the time cost when using HLF, while for a block that has been transmitted to the remote DB server, the query latency is similar to that when using xFabLedger.

7. Conclusions

In this paper, a secured and accelerated scalable storage solution for blockchain systems is proposed. Our solution relieves the storage burden of blockchain nodes by adding an off-chain remote DB server for each CU in blockchain network to store the blocks while ensuring data integrity. Our solution also improves the performance compared to our previous work by separating the data-deduction operation from the transaction-processing phase. Experiments show that our solution can reduce the size of block files for peers by 93.3% compared to the original HLF blockchain system and that our solution enhances the TPS by 9.6% compared to xFabLedger.

Our solution can be applied in blockchain systems constructed with HLF to enhance its storage scalability by simply replacing the vanilla HLF peers with the modified SASLedger peers and deploying the remote DB server. For conventional HLF blockchain systems, if the storage spaces of peers are insufficient, additional hard disks need to be installed to each peer; however, in a SASLedger blockchain system, new hard disks only need to be installed in the remote DB server, which reduces the budget.

There are several limitations left in our solution:

- Retrieving blocks or transactions from the remote DB server is slower than retrieving them from the local file system of a peer, thus our solution leads to a decrease of the query performance.
- The data-reduction phase of our solution depends on the snapshot operation of HLF, since the commitment of transactions are stopped when taking the snapshot, the performance is decreased due to the snapshot operation.

In the future, we would like to study the mechanism of the local databases of HLF peers to solve the read-write conflict problem when exporting the contents of the local databases. This may accelerate the current snapshot generation process by keeping the commitment of transactions unstopped when a snapshot is taken.

Author Contributions: Conceptualization, B.P., J.S. and H.S.; methodology, H.S. and B.P.; software, H.S.; validation, H.S. and B.P.; formal analysis, B.P.; investigation, H.S. and B.P.; resources, H.S. and B.P.; data curation, B.P.; writing—original draft preparation, H.S.; writing—review and editing, B.P., J.S., T.M. and M.M.; visualization, H.S.; supervision, J.S.; project administration, B.P. All authors have read and agreed to the published version of the manuscript.

Funding: This research received no external funding.

Data Availability Statement: Not Applicable, the study does not report any data.

Conflicts of Interest: The authors declare no conflict of interest.

Abbreviations

Abbreviations
The following abbreviations are used in this manuscript:

DB	Database
HLF	Hyperledger Fabric
Org	Organization
OSN	Ordering Service Nodes
BlkNum	Block Number
BlkHash	Block Hash
Tx	Transaction
PBFT	Practical Byzantine Fault Tolerant
VC	Vector Commitment
CU	Consensus Unit
BAO	Blocks Assignment Optimization
TPS	Transactions Per Second
OS	Operating System
CPU	Central Processing Unit
SSD	Solid-State Driver
GB	Gigabytes
MB	Megabytes

Appendix A.

In the following algorithms, "=" means "equals to" and "||" means "concatenation".

Algorithm A1: Initialization

$txArray \leftarrow [tx_1, ..., tx_n]$
$MerkleTree.layerArray \leftarrow emptyArray$ /* Initialize the layers of the MerkleTree with an empty array */

Function CalcHashForTxs(*txArray*):
  $i \leftarrow 0$
  while $i < sizeOf(txArray)$ **do**
    $hashArray[i] \leftarrow sha256(txArray[i] \;||\; i)$
    $i \leftarrow i + 1$
  end
  return *hashArray*

Algorithm A2: Commitment and Construction of MerkleTree

Function Commitment(*txArray*):
  hashArray ← *CalcHashForTxs*(*txArray*)
  return *CommitRecursively*(*hashArray*)

Function CommitRecursively(*layer*):
  $i \leftarrow 0$
  defaultSibling ← *sha256*(*constantA*)
  while $i <$ *sizeOf(layer)* **do**
    if *layer*[$i+1$] *exists* **then**
      if *layer[i] < layer[i+1]* **then**
        | *upperLayer*[$\frac{i}{2}$] ← *sha256*(*layer*[i] || *layer*[$i+1$])
      else
        | *upperLayer*[$\frac{i}{2}$] ← *sha256*(*layer*[$i+1$] || *layer*[i])
      end
      layer[i].*parent* ← *upperLayer*[$\frac{i}{2}$]   /* Record the node's parent */
      layer[i].*sibling* ← *layer*[$i+1$]   /* Record the node's sibling */
      layer[$i+1$].*parent* ← *upperLayer*[$\frac{i}{2}$]
      layer[$i+1$].*sibling* ← *layer*[i]
    else
      if *layer[i] < defaultSibling* **then**
        | *upperLayer*[$\frac{i}{2}$] ← *sha256*(*layer*[i] || *defaultSibling*)
      else
        | *upperLayer*[$\frac{i}{2}$] ← *sha256*(*defaultSibling* || *layer*[i])
      end
      layer[i].*parent* ← *upperLayer*[$\frac{i}{2}$]
      layer[i].*sibling* ← *defaultSibling*
    end
    $i \leftarrow i + 2$
  end
  MerkleTree.layerArray.append(*layer*)   /* Save current layer to MerkleTree */
  if $1 = sizeOf(upperLayer)$ **then**
    MerkleTree.layerArray.append(*upperLayer*)   /* Save the upper layer to MerkleTree */
    return *upperLayer*[0]
  else
    | **return** *CommitRecursively*(*upperLayer*)
  end

Algorithm A3: Open

Function Open($i, txArray$):
  $hashArray \leftarrow CalcHashForTxs(txArray)$
  $proofArray \leftarrow emptyArray$
  return $GetProofRecursive(hashArray[i], proofArray)$

Function GetProofRecursively($node, proofArray$):
  $sibling \leftarrow MerkelTree.getSibling(node)$ /* Find the node's sibling */
  if *sibling dose not exist* **then**
    return $proofArray$
  else
    $proofArray.append(sibling)$
    $parent \leftarrow MerkelTree.getParent(node)$ /* Find the node's parent */
    return $GetProofRecursively(parent, proofArray)$
  end

Algorithm A4: Verify

Function Verify($C, i, tx, proofArray$):
  $result \leftarrow sha256(tx \,||\, i)$
  $j \leftarrow 0$
  while $j < sizeOf(proofArray)$ **do**
    if *result < proofArray[j]* **then**
      $result \leftarrow hash256(result \,||\, proofArray[j])$
    else
      $result \leftarrow hash256(proofArray[j] \,||\, result)$
    end
    $j \leftarrow j + 1$
  end
  if $C = result$ **then**
    return 1
  else
    return 0
  end

References

1. Nakamoto, S. Bitcoin: A Peer-to-Peer Electronic Cash System. Available online: https://bitcoin.org/bitcoin.pdf (accessed on 15 October 2021).
2. Ethereum Home. Available online: https://ethereum.org/en/ (accessed on 15 October 2021).
3. Home—EOSIO Blockchain Software & Services. Available online: https://eos.io/ (accessed on 5 November 2021).
4. Androulaki, E.; Barger, A.; Bortnikov, V.; Cachin, C.; Christidis, K.; De Caro, A.; Enyeart, D.; Ferris, C.; Laventman, G.; Manevich, Y.; et al. Hyperledger fabric: A distributed operating system for permissioned blockchains. In Proceedings of the Thirteenth EuroSys Conference, Porto, Portugal, 23–26 April 2018; Association for Computing Machinery: New York, NY, USA, 2018; pp. 1–15.
5. Quorum. Available online: https://consensys.net/quorum/ (accessed on 8 October 2021).
6. Bitcoin Blockchain Size. Available online: https://ycharts.com/indicators/bitcoin_blockchain_size (accessed on 29 September 2021).
7. Ethereum Chain Full Sync Data Size. Available online: https://ycharts.com/indicators/ethereum_chain_full_sync_data_size (accessed on 29 September 2021).
8. 10 Practical Issues for Blockchain Implementations. Available online: https://www.hyperledger.org/blog/2020/03/31/title-10-practical-issues-for-blockchain-implementations (accessed on 24 November 2021).
9. Ledger Snapshot and Checkpoint. Available online: https://jira.hyperledger.org/browse/FAB-106 (accessed on 24 November 2021).

10. Chow, S.S.; Lai, Z.; Liu, C.; Lo, E.; Zhao, Y. Sharding blockchain. In Proceedings of the 2018 IEEE International Conference on Internet of Things (iThings) and IEEE Green Computing and Communications (GreenCom) and IEEE Cyber, Physical and Social Computing (CPSCom) and IEEE Smart Data (SmartData), Halifax, NS, Canada, 30 July–3 August 2018; p. 1665.
11. Wang, J.; Wang, H. Monoxide: Scale out blockchains with asynchronous consensus zones. In Proceedings of the 16th USENIX Symposium on Networked Systems Design and Implementation (NSDI), Boston, MA, USA, 26–28 February 2019; pp. 95–112.
12. Xu, Z.; Han, S.; Chen, L. CUB, a consensus unit-based storage scheme for blockchain system. In Proceedings of the 2018 IEEE 34th International Conference on Data Engineering (ICDE), Paris, France, 16–19 April 2018; pp. 173–184.
13. Dai, X.; Xiao, J.; Yang, W.; Wang, C.; Jin, H. Jidar: A jigsaw-like data reduction approach without trust assumptions for bitcoin system. In Proceedings of the 2019 IEEE 39th International Conference on Distributed Computing Systems (ICDCS), Dallas, TX, USA, 7–10 July 2019; pp. 1317–1326.
14. Honar Pajooh, H.; Rashid, M.; Alam, F.; Demidenko, S. Hyperledger Fabric Blockchain forSecuring the Edge Internet of Things. *Sensors* **2021**, *21*, 359. [CrossRef] [PubMed]
15. Hwang, H.C.; Shon, J.G.; Park, J.S. Design of an Enhanced Web Archiving System for Preserving Content Integrity with Blockchain. *Electronics* **2020**, *9*, 1255. [CrossRef]
16. Galiev, A.; Prokopyev, N.; Ishmukhametov, S.; Stolov, E.; Latypov, R.; Vlasov, I. Archain: A novel blockchain based archival system. In Proceedings of the 2018 Second World Conference on Smart Trends in Systems, Security and Sustainability (WorldS4), London, UK, 30–31 October 2018; pp. 84–89.
17. Miyamae, T.; Kozakura, F.; Nakamura, M.; Zhang, S.; Hua, S.; Pi, B.; Morinaga, M. ZGridBC: Zero-Knowledge Proof based Scalable and Private Blockchain Platform for Smart Grid. In Proceedings of the 2021 IEEE International Conference on Blockchain and Cryptocurrency (ICBC), Sydney, Australia, 3–6 May 2021; pp. 1–3.
18. Hyperledger Fabric Block Archiving. Available online: https://github.com/hyperledger-labs/fabric-block-archiving (accessed on 18 October 2021)
19. Catalano, D.; Fiore, D. Vector commitments and their applications. In Proceedings of the International Workshop on Public Key Cryptography (PKC), Nara, Japan, 26 February–1 March 2013; Springer: Berlin/Heidelberg, Germany, 2013; pp. 55–72.
20. Pi, B.; Pan, Y.; Zhou, E.; Sun, J.; Miyamae, T.; Morinaga, M. xFabLedger: Extensible Ledger Storage for Hyperledger Fabric. In Proceedings of the 2021 IEEE 11th International Conference on Electronics Information and Emergency Communication (ICEIEC), Beijing, China, 18–20 June 2021; pp. 5–11.
21. A Blockchain Platform for the Enterprise. Available online: https://hyperledger-fabric.readthedocs.io/en/release-2.3/index.html (accessed on 8 October 2021).
22. Chen, B.; Tan, Z.; Fang, W. Blockchain-based implementation for financial product management. In Proceedings of the 2018 28th International Telecommunication Networks and Applications Conference (ITNAC), Sydney, Australia, 21–23 November 2018; pp. 1–3.
23. Chen, Y.; Bellavitis, C. Blockchain disruption and decentralized finance: The rise of decentralized business models. *J. Bus. Ventur. Insights* **2020**, *13*, e00151. [CrossRef]
24. Caldarelli, G.; Ellul, J. The Blockchain Oracle Problem in Decentralized Finance—A Multivocal Approach. *Appl. Sci.* **2021**, *11*, 7572. [CrossRef]
25. Sun, H.; Hua, S.; Zhou, E.; Pi, B.; Sun, J.; Yamashita, K. Using ethereum blockchain in Internet of Things: A solution for electric vehicle battery refueling. In Proceedings of the International Conference on Blockchain, Seattle, WA, USA, 25–30 June 2018; pp. 3–17.
26. Xu, R.; Nagothu, D.; Chen, Y. EconLedger: A Proof-of-ENF Consensus Based Lightweight Distributed Ledger for IoVT Networks. *Future Internet* **2021**, *13*, 248. [CrossRef]
27. Połap, D.; Srivastava, G.; Yu, K. Agent architecture of an intelligent medical system based on federated learning and blockchain technology. *J. Inf. Secur. Appl.* **2021**, *58*, 102748. [CrossRef]
28. Wang, B.; Li, Z. Healthchain: A Privacy Protection System for Medical Data Based on Blockchain. *Future Internet* **2021**, *13*, 247. [CrossRef]
29. Yaqoob, I.; Salah, K.; Jayaraman, R.; Al-Hammadi, Y. Blockchain for healthcare data management: Opportunities, challenges, and future recommendations. *Neural Comput. Appl.* **2021**, 1–16. [CrossRef]
30. Zhou, Q.; Huang, H.; Zheng, Z.; Bian, J. Solutions to scalability of blockchain: A survey. *IEEE Access* **2020**, *8*, 16440–16455. [CrossRef]
31. Khan, D.; Jung, L.T.; Hashmani, M.A. Systematic Literature Review of Challenges in Blockchain Scalability. *Appl. Sci.* **2021**, *11*, 9372. [CrossRef]
32. Antal, C.; Cioara, T.; Anghel, I.; Antal, M.; Salomie, I. Distributed Ledger Technology Review and Decentralized Applications Development Guidelines. *Future Internet* **2021**, *13*, 62. [CrossRef]
33. Stoica, I.; Morris, R.; Karger, D.; Kaashoek, M.F.; Balakrishnan, H. Chord: A scalable peer-to-peer lookup service for internet applications. *ACM Sigcomm Comput. Commun. Rev.* **2001**, *31*, 149–160. [CrossRef]
34. Maymounkov, P.; Mazieres, D. Kademlia: A peer-to-peer information system based on the xor metric. In Proceedings of the International Workshop on Peer-to-Peer Systems, Cambridge, MA, USA, 7–8 March 2002; Springer: Berlin/Heidelberg, Germany, 2002; pp. 53–65.
35. Flooding. Available online: https://www.cs.yale.edu/homes/aspnes/pinewiki/Flooding.html (accessed on 21 October 2021).

36. Gorenflo, C.; Lee, S.; Golab, L.; Keshav, S. FastFabric: Scaling hyperledger fabric to 20 000 transactions per second. *Int. J. Netw. Manag.* **2020**, *30*, e2099. [CrossRef]
37. Blockchain Forks Explained. Available online: https://medium.com/digitalassetresearch/blockchain-forks-explained-8ccf304b97c8 (accessed on 21 October 2021).
38. The Raft Consensus Algorithm. Available online: https://raft.github.io/ (accessed on 21 October 2021).
39. Castro, M.; Liskov, B. Practical byzantine fault tolerance. In Proceedings of the Third Symposium on Operating Systems Design and Implementation (OSDI), New Orleans, LA, USA, 22–25 February 1999; Volume 99, pp. 173–186.
40. CouchDB. Available online: http://couchdb.apache.org/ (accessed on 21 October 2021).
41. Google/Leveldb. Available online: https://github.com/google/leveldb (accessed on 21 October 2021).
42. Taking Ledger Snapshots and Using Them to Join Channels. Available online: https://hyperledger-fabric.readthedocs.io/en/release-2.3/peer_ledger_snapshot.html (accessed on 21 October 2021).
43. Merkle, R.C. A digital signature based on a conventional encryption function. In Proceedings of the Conference on the Theory and Application of Cryptographic Techniques, Santa Barbara, CA, USA, 16–20 August 1987; Springer: Berlin/Heidelberg, Germany, 1987; pp. 369–378.
44. gRPC. Available online: https://grpc.io/ (accessed on 21 October 2021).
45. Performance Traffic Engine—PTE. Available online: https://github.com/hyperledger/fabric-test/tree/main/tools/PTE (accessed on 21 October 2021).

Article

Distributed Hybrid Double-Spending Attack Prevention Mechanism for Proof-of-Work and Proof-of-Stake Blockchain Consensuses

Nur Arifin Akbar [1], Amgad Muneer [2], Narmine ElHakim [3] and Suliman Mohamed Fati [3,*]

[1] Research Department, Idenitive Mashable Prototyping, Banyumas 53124, Indonesia; arifin@idenitive.pro
[2] Department of Computer and Information Sciences, Universiti Teknologi PETRONAS, Seri Iskandar 32160, Malaysia; muneeramgad@gmail.com
[3] College of Computer and Information Sciences, Prince Sultan University, Riyadh 11586, Saudi Arabia; Nhakim@psu.edu.sa
* Correspondence: smfati@yahoo.com or sgaber@psu.edu.sa

Citation: Akbar, N.A.; Muneer, A.; ElHakim, N.; Fati, S.M. Distributed Hybrid Double-Spending Attack Prevention Mechanism for Proof-of-Work and Proof-of-Stake Blockchain Consensuses. *Future Internet* **2021**, *13*, 285. https://doi.org/10.3390/fi13110285

Academic Editor: Ahad ZareRavasan

Received: 9 October 2021
Accepted: 4 November 2021
Published: 12 November 2021

Publisher's Note: MDPI stays neutral with regard to jurisdictional claims in published maps and institutional affiliations.

Copyright: © 2021 by the authors. Licensee MDPI, Basel, Switzerland. This article is an open access article distributed under the terms and conditions of the Creative Commons Attribution (CC BY) license (https://creativecommons.org/licenses/by/4.0/).

Abstract: Blockchain technology is a sustainable technology that offers a high level of security for many industrial applications. Blockchain has numerous benefits, such as decentralisation, immutability and tamper-proofing. Blockchain is composed of two processes, namely, mining (the process of adding a new block or transaction to the global public ledger created by the previous block) and validation (the process of validating the new block added). Several consensus protocols have been introduced to validate blockchain transactions, Proof-of-Work (PoW) and Proof-of-Stake (PoS), which are crucial to cryptocurrencies, such as Bitcoin. However, these consensus protocols are vulnerable to double-spending attacks. Amongst these attacks, the 51% attack is the most prominent because it involves forking a blockchain to conduct double spending. Many attempts have been made to solve this issue, and examples include delayed proof-of-work (PoW) and several Byzantine fault tolerance mechanisms. These attempts, however, suffer from delay issues and unsorted block sequences. This study proposes a hybrid algorithm that combines PoS and PoW mechanisms to provide a fair mining reward to the miner/validator by conducting forking to combine PoW and PoS consensuses. As demonstrated by the experimental results, the proposed algorithm can reduce the possibility of intruders performing double mining because it requires achieving 100% dominance in the network, which is impossible.

Keywords: blockchain; proof of work; proof of stake; consensus mechanism; 51% attack; double-mining attack; technological development

1. Introduction

Blockchain technology has been widely used in various distributed system contexts, including content distribution networks [1], smart grid systems [2], e-healthcare [3], real estate [4,5], e-finance [6], e-education [7], supply chains, e-voting, smart homes [8,9], smart cities [10] and smart industries [11,12]. The advent of blockchain technology has affected the global financial system through digital currencies. In 2008, Satoshi Nakamoto invented a revolutionary electronic cash system called Bitcoin (a digital currency) that made peer-to-peer electronic transactions possible. This peer-to-peer digital currency system was designed to eliminate the need for third parties in financial transactions between unknown parties in a trustworthy and verifiable way [13]. In January 2009, the same group created software as an open-source code and introduced the first digital currency in history [14]. As the fundamental technology of Bitcoin, blockchain consists of a transparent and immutable list of chained blocks of transactions. In the peer-to-peer network, each peer maintains a copy of the blockchain known as the distributed ledger.

Blockchain acts as a decentralised public ledger for recording data as blocks, which constitute a connected list data structure used to indicate logical relationships between the data added to the blockchain. The data blocks can be retained without the involvement of a centralised agency or intermediary. In another alternative, data blocks are copied and exchanged throughout the entire blockchain network, thereby eliminating device failure, data management and cyber-attacks. The two most important processes of blockchain are block mining and block validation. The mining process involves adding a new block or transaction to the public global ledger. The new block or transaction is then validated in a process known as block validation. To understand how blockchain operates, we need to understand its four underlying layers. At the lowest layer are peers sign transactions, which represent an agreement between two parties, such as exchanging physical or digital property or completing a task. To ensure the absence of corrupt branches and divergences [15], the nodes must agree on which transactions should be kept in the blockchain, which is the responsibility of the consensus layer. The third layer is the compute interface. Through the compute interface, the blockchain is able to provide increased functionality. Blockchain maintains a record of each transaction undertaken by a user so that by calculating the balance of each user, the overall balance may be determined. The last layer, governance, extends the blockchain architecture to human interaction in the physical realm. Therefore, the popularity of blockchain is inevitable because the technology can provide desirable features by replacing the centralised communication architectures of today. The core protocol of blockchain, particularly in blockchain-based cryptocurrencies, refers to the consensus protocol. The consensus protocol enables all peers to agree on every block inclusion in the distributed ledger [16]. As a result of a consensus mechanism, all truthful nodes establish mutual agreement on a consistent ledger in asynchronous, untrusted networks [17]. The consensus protocols are well-defined, but inputs from various stakeholders are also considered, which affects the blockchain's authenticity. Incorporating new methods for improving consensus protocols and/or patching systems is therefore essential to the development of blockchains.

Different consensus mechanisms are required to ensure the security of digital transactions due to the varying types of blockchain technology [18]. A common consensus mechanism is proof-of-work (PoW), in which the parties must demonstrate their rights to add a node by solving an increasingly complicated computational problem to ensure authentication and compliance, including identifying thresholds for harm, such as leading zeros [19]. Given that the PoW protocol needs tremendous computing power to solve the block complexity in Bitcoin [20], another consensus protocol called proof-of-stake (PoS) was proposed to overcome the problems of the PoW protocol. Despite the high complexity of the PoS consensus, this protocol may be vulnerable to stack problems if more than half of the network is manipulated to prevent a new block from being distributed to confirm transactions [21]. A PoS protocol separates stake blocks according to the relative hashing rates of miners (i.e., their computational power) in relation to the resource capacity of existing miners [22]. This approach makes the choice fair and prevents the richest participant from dominating the network. Many blockchains, such as Ethereum [23], opt for PoS because power consumption and scalability are greatly reduced. Several consensus approaches, including Byzantine fault tolerance (BFT) and its variants, are also available [24].

However, despite the application of consensus protocols, which prevent many security breaches, several malicious attacks have occasionally hampered the growth of blockchain technology. For example, certain attacks, such as Eclipse, Sybil, BGP deterrence, and 51%, are triggered as a result of attempts to penetrate the blockchain network. Amongst these attacks, the 51% attack has received the least attention from researchers due to its high costs. However, recent security incidents have demonstrated that 51% attacks can be carried out against various contemporary cryptocurrencies [25]. Compared with other consensus protocols, PoW immediately challenges 51% attacks, where recent attacks have mainly focused on PoW-dependent cryptocurrencies [26]. This is one of the most severe dangers associated with a PoW-based cryptocurrency because it assumes that if a fraudulent peer network is allowed to obtain more than 50% of the network assets (i.e., computing power), its members become the majority of the network's decision makers. Peers with superior processing skills could dominate the network because they have the capability to mine numerous blocks as peers compete for fast access. They can easily exploit the blockchain by creating fake transactions, and the fraud perpetrated by other users may result in large-scale financial losses.

To prevent this attack, researchers have performed various studies. The majority of them recommended combining two or more resource proofs into a hybrid protocol to combat this attack [27–31]. However, mixing two or more existing protocols (hybrid protocol) makes the network resistant to this attack. Therefore, the recent implementation of hybrid protocols has other challenges and drawbacks that need to be addressed. For example, several have added voting systems, ticket delivery systems, fines, special nodes and block validator groups to deter malicious behaviour [32]. These measures are successful in protecting the network against 51% attacks. However, their primary weakness is in rewarding block mining to investors, which pertains to the number of Bitcoins you receive if you are successful in mining a block. Undoubtedly, the investor invests his hard-earned money in a cryptocurrency to reap the benefits of his investment. These benefits may be derived from the block mining reward. In this scenario, the accuracy of the block generation time interval is crucial in ensuring that this benefit is delivered to the appropriate consumer at the appropriate time. However, the voting, ticket and other systems are not time-controlled, and no consistent distribution of benefits occurs over the block reward generation intervals. Another major issue is the diversification of peers by establishing special committees and validation groups that violate the P2P network's principle.

Hence, this study proposes a hybrid consensus protocol that integrates PoW and PoS to control block generation time in two ways. Firstly, our proposed model uses the PoW mining method for the first time to prevent the block generation time from exceeding a specified threshold. Secondly, the generated block is validated by the PoS consensus without any need for voting or commission approval. In the proposed model, each block is validated by the entire network. Hybridisation is one of the aspects that make our study unique and novel compared with previous studies. In addition to being able to handle the 51% attack, the framework ensures a standardised distribution of mining rewards to stakeholders and investors by maintaining a precise block generation interval with difficulty adjustment in PoW mining and stakeholder probability calculation based on their mature stake balance. This study proposes a hybrid algorithm that combines the PoW and PoS mechanisms to ensure a fair mining reward between the miner and validator by controlling the block generation time. To ensure long-term sustainability, the proposed model entails a complexity analysis. The important contributions of this work can be summarised as follows:

- We evaluated three security protection measures that are specific to the 51% attack and demonstrated their vulnerabilities to exploitation by the 51% attack.
- We proposed a model to control the block generation time with the distributed validation technique, which enhances blockchain security and performance.
- We hybridised PoW and PoS consensuses to solve the above-mentioned issue for the fair mining and stacking mechanism, which by default prevents the 51% attack.

This paper is structured as follows. Section 2 presents a background of the topic and related work wherein blockchain and previous attempts are described and investigated. Section 3 provides an overview of the methodology adopted in this study and a description of the experiment's algorithms. The analysis and results are given in Section 4, and the conclusions and future work directions are presented in Section 5.

2. Background and Related Work

In the past few years, blockchain technology has been applied to cryptocurrencies. In the blockchain concept, data are exchanged from peer to peer in a distributed and decentralised manner [33]. In principle, all blockchain technologies employ the concept of a distributed ledger, in which the data are stored on a decentralised mechanism, with a cryptographic key being distributed across the network to ensure that each transaction matches its corresponding entity [34]. Data should be checked and validated before entering the ledger, which is why a consensus protocol is required. Although the scalability trilemma affects the development of blockchain technologies, this trilemma has no formal definition in the literature, but it has been reported in numerous studies, such as in [35,36]. Its definition was coined by Vitalik Buterin, the developer of Ethereum, a blockchain based on PoW that specifies the three characteristics a blockchain must possess if it is to expand globally: decentralisation, security and scalability. As shown in Figure 1, the scalability trilemma is symbolised by a triangle whose vertices represent three properties. The blockchain with optimal scalability is at the centre of the figure, which is not currently applicable. Brief descriptions of the three properties are presented below.

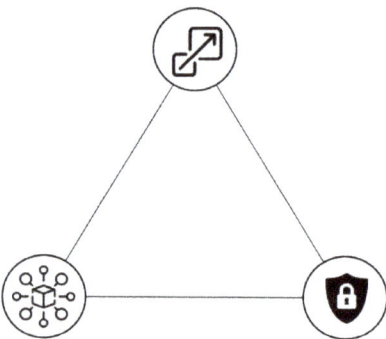

Figure 1. Representation of the blockchain scalability trilemma; according to the trilemma, a blockchain can be on one side of the triangle and not in the center, which represents the best.

2.1. Decentralisation

Decentralisation ensures that transactions are verified and confirmed by a community of nodes and not by a central authority or a select committee, as in conventional systems. In other words, decisions are made by a distributed consensus, so any transaction does not require the trust of a third party. Consequently, the decisions made by network members are democratic, and any changes made to the protocol will be approved if more than 50% of the participants agree. An example is the fork in Bitcoin Cash that occurred on 1 August 2017 [37], where the maximum size of a block was increased to 8 MB to allow more transactions to be accepted. Given that multiple nodes verify the decision, decentralisation leads to higher-quality decisions than centralised authorities. The trade-off is the speed of confirmation; if a transaction requires the confirmation of multiple participants, the speed is less than that of a decision made by a central authority [38].

2.2. Scalability

Global adoption is enabled by the property of scalability, which refers to a system's capability to adjust to increased loads. Bitcoin and Ethereum, two of the most widely used blockchain technologies, can process a maximum of seven and twelve transactions per second (TPS) unlike Visa, which can process 65,000 TPS [39]. Moreover, EOS [40], which is designed to be scalable, claims a throughput of around 2000 TPS but promises to be able to process millions of transactions in the future at the price of decentralisation.

2.3. Security

Security is a fundamental requirement in a blockchain. Insufficient or absent security permits an attacker to spend the same amount several times (double spending), thereby enriching himself at the expense of others and changing the blockchain's immutable status. Such a scenario could occur in a 51% attack. Notably, the blockchain scalability trilemma is not a theorem, but in the context of distributed systems, the combination of consistency, availability and partition tolerance (CAP) is a fundamental theorem. This combination emphasises the difficulty of creating a decentralised, secure, scalable system, particularly in the case of blockchain technology, which is still developing and immature. The Bitcoin Blockchain, for example, features high security and decentralisation, but it is not scalable; the maximum number of transactions it can support is seven per second. Although it is not used as the sole currency, it represents a significant milestone in computer history by demonstrating the use of a digital cryptocurrency in a peer-to-peer network.

Additionally, most public blockchains, such as Bitcoin and Ethereum Classic [41], use a PoW consensus protocol to ensure that the data are immutable because all transactions must be mined to solve the complexity of the block. However, this protocol is susceptible to double-spending attacks, which occur when a user makes a second transaction with the same data as a previous one that has already been validated. Furthermore, if the miner controls more than 50% of the computing power managing the blockchain, he might be able to prevent the generation of a new block because any proposed change to the protocol must be supported by more than 50% of the participants. Therefore, amongst the numerous attacks that affect Blockchain protocols, 51% should receive additional attention. As a rule of thumb, blockchain technology is based on a distributed consensus mechanism that ensures mutual trust. When a miner owns more than 50% of the hash power in a PoW-based blockchain, he can carry out a 51% attack. In this case, he will receive 100% of the rewards from mining because he will create blockchains that are longer than those of any other miner. A double-spending attack can also occur if the same unspent transaction output (UTXO) is used for two transactions at the same time, thus erasing the last confirmed blocks from the blockchain and possibly corrupting the blockchain itself. Throughout the years, technologies such as Bitcoin that economically incentivise nodes to become miners have increased to a high number of nodes. Therefore, such an attack would require a considerable amount of hash power. Small blockchains, which have a hash power that is lower than that of Bitcoin, are not excluded from this attack. Examples of cryptocurrencies that are affected by 51% attacks include Monacoin [42], Bitcoin Gold [43] and ZenCash [44]. Furthermore, mining pools entail several miners sharing their computational power with several others who share the compensation proportionately to their shares of computing power. Owing to the advent of Bitcoin mining pools, an organisation can carry out a 51% attack if the sum of the hash power of all registered nodes exceeds 50% of the total network hash power. An example of Bitcoin blockchain 51% attack is shown in Figure 2.

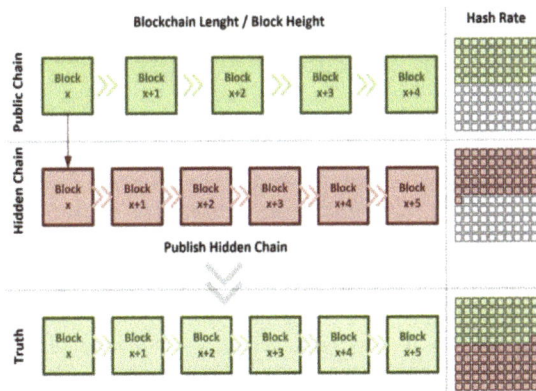

Figure 2. Illustration of Bitcoin blockchain 51% attack. If an attacker acquires more than half of the global hashing power, they will be able to mine a hidden chain that will eventually surpass the length of the public chain. Once the hidden chain surpasses the length of the public chain, it can be published and accepted as the new truth.

In addition, an attack with new hash power implies that an attacker has opted to obtain a more powerful hash power than that of the public chain, having the advantage of not knowing the start of the attack because no hashing power will leave the live network. Apart from the fact that this is a stealth attack, it does not force the difficulty to adjust to the live network as a result of a drop in its hash rate, thereby preventing new miners from joining the live network. In this case, the only factor that affects the live chain's hash rate and complexity is the Bitcoin price itself. By contrast, this attack is twice as costly to execute as the current hash power attack. The execution of a 51% attack with new hash power is shown in Figure 3.

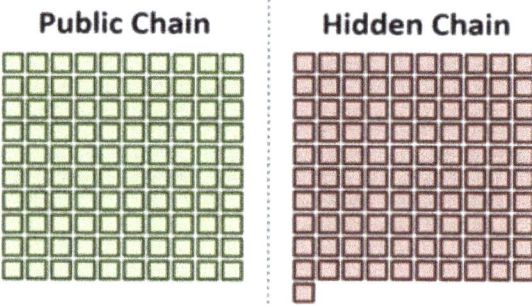

Figure 3. Execution of a 51% attack with new hash power.

In sum, according to the literature, current systems have different aspects that make blockchains non-scalable [45]. Three critical aspects stand out in particular:

- New transactions are sent in broadcast to all the nodes of the network.
- All nodes receive each new block.
- A set of nodes is responsible for processing all the new transactions involved in the next block.

Given these aspects, the computational capacity and bandwidth of the individual nodes must be proportionally increased (vertical scalability). Horizontal scalability, in which additional nodes are added to the network in response to the increased volume of transactions, is preferable.

2.4. Consensus Mechanisms

In the blockchain network, the consensus mechanism is a set of rules designed to guarantee that all participants adhere to the same set of rules. The protocol ensures that each participant's consent is used to carry out transactions to the distributed ledger [39]. A public blockchain is a decentralised technology, and no central authority is responsible for governing the necessary action. For this reason, the blockchain network requires the permission of network participants to verify and authenticate the activities taking place in the network. The entire process is executed by consensus amongst network members, which makes blockchain a trustworthy, secure, and efficient technology for digital transactions. Different consensus frameworks follow different standards that allow participants in the network to comply with these rules. To address the concerns of safe digital transactions, several consensus processes have been implemented. A few consensus protocols employed by major cryptocurrencies are PoW, PoS and delegated PoS (DPoS).

2.4.1. PoW

During mining, a new block is created by computing the block's cryptographic hash. To prove its validity on the blockchain, a block hash must meet certain conditions. The

Bitcoin blockchain, for example, starts each hash block with four trailing zeroes. Given that block data, which are transactional data, cannot be changed, the miner must modify the predefined hash pattern at every occurrence. The two network partners compete for the right nonce to create a valid block hash. Initially, the miner who seeks a solution attaches the block to the chain. As a reward for the miner's efforts, the system produces a certain number of coins and provides the newly produced coins to the miner.

The PoW mechanism entirely relies on the computer power of the miner. The more computing power a miner has, the greater the chance of finding blocks and earning rewards [46]. In PoW consensus, half of the network's nodes are assumed to remain trustworthy. As a result, this consensus is vulnerable because more than half the hashing power is owned by a single party. The cost of resources and hardware is one of the significant disadvantages of PoW. Several studies have reported that the energy consumption of Bitcoin mining is considerably higher than the energy consumption of 159 countries [47]. By contrast, the mining requirements and the mining time can differ depending on the algorithm used by each cryptocurrency. PoW mining is relatively slower than other consensus protocols. Given that a small number of mining pools dominate the Bitcoin network, attacks on these pools may result in severe disruptions. Recent attacks have demonstrated that PoW is vulnerable to 51% attacks. Low-hacking crypto coins based on PoW consensus are susceptible to 51% attacks because the requisite hash is easy to obtain. With the appropriate budget, the P+ epsilon attack can be conducted at no cost [48]. In addition, the researchers in [49] studied blockchain security and performance-based PoW. They presented a novel quantitative approach to examine the security and performance implications of various consensus and network parameters applied to PoW blockchains. Therefore, the approach proposed in [49] is solely based on PoW consensus, as opposed to hybrid mechanism approaches that offer higher levels of security and performance. Another study was conducted by [50] to review the role of blockchain in preventing future pandemics. Several applications of blockchain technology were also discussed, and these may assist in fighting the COVID-19 pandemic.

2.4.2. PoS

PoS is not dependent on a high computation capacity. It operates according to the staked properties of the network. In general, the more money a peer receives, the higher the incentive they will have to mine and reward. This mechanism does not require extremely high computational power. Little calculating power is required, so excessive electricity consumption is reduced [27]. Several limitations are associated with the PoS protocol. For example, large investors with enormous capital can manage the network to maximise their wealth, making the rich richer. PoS is also vulnerable to 51% attacks, especially when someone has more than 50% of the network wealth. An individual can exploit the blockchain easily for personal gain, resulting in malicious stakeholders gaining the majority of the supply by taking advantage of the nothing-at-stake issue. PoS suffers from low subjectivity and is thus challenging and demanding to implement [51]. To conduct a 51% attack, an opponent must obtain 51% of all cryptocurrencies. The cost of obtaining 51% of the overall stake is substantial. In light of this, the threat level posed by the 51% attack may be lower than that posed by PoW. According to our analysis, a long-range attack can exploit PoS. Carrying out the P + Epsilon attack is impossible because a large budget is required for an attacker to donate to the minority's safety deposit. A Sybil attack can exploit PoS, and a DPoS attack can interrupt any portion of the network.

2.4.3. DPoS

DPoS is a consensus process that allows shareholders to vote on the nomination of witnesses [52]. In DPoS, the main objective is to minimise energy waste and accelerate transaction times. As a result of the overall block generation process, this consensus mechanism operates much more quickly than PoW consensus. In DPoS, each stakeholder is allowed to cast one vote per share; they can cast additional votes when they own additional coins. Moreover, the witnesses are rewarded for producing blocks and penalised for failing to do so, such that they are not paid and are voted out of office. To complete the instructed task, witnesses should receive the largest number of votes from random stakeholders. A stakeholder also votes on the restructuring of the delegates and adjusts the network, which will be reviewed by the stakeholder before a final decision. Although DPoS was designed to increase transaction efficiency and overcome the constraints imposed by many other consensus mechanisms, it has significant shortcomings. The network is not sufficiently decentralised due to a large number of validators. A centralised system may serve as a focal point for random intruders due to its centralised nature. DPoS is susceptible to 51% attacks because an attacker will convince stakeholders to give them 51% voting power in a 51% attack [40]. Asymmetric agreements are also vulnerable to other types of attacks, such as long-distance, DDoS, P + epsilon, Sybil and balanced attacks.

This work investigated the fact that the three main consensus systems, which are susceptible to several attacks, have significant weaknesses. As a result of their vulnerability, digital transactions are at a high risk of being attacked. Table 1 summarises the results of our analysis. The 51% attack can exploit all three consensus mechanisms, making it desirable for attackers, particularly for PoW where achieving the required hashing power is cost-effective.

Table 1. Vulnerabilities of consensus mechanisms.

Consensus Mechanism	51% Attack	Long-Range Attack	DDoS Attack	P + Epsilon Attack	Sybil Attack	Balance Attack	BGP Hijacking
PoW	Yes	No	Yes	Yes	Yes	Yes	Yes
PoS	Yes	Yes	Yes	No	Yes	No	No
DPoS	Yes	Yes	Yes	Yes	Yes	Yes	No

The following is a brief description of several severe attacks. However, the focus of this study is on 51% attacks. Long-range attacks are the result of a weak model of subjectivity [53]. This form of attack is similar to the 51% attack. It appears to fork the chain from the genesis block [54] rather than confirming the sixth block. This type of attack occurs very rarely in Bitcoin, but it can be damaging when demonstrating stakeholder consensus (PoS) and delegate stakeholder consensus (DPoS). Assuming a PoS consensus scenario in which the invaders begin with a limited number of coins shortly after the genesis block, their chain versions can be privately mined to carry out the attack. Given that they have a small stake, they will generate a limited number of blocks at the beginning and then generate a longer chain. PoS does not specify a threshold for chain lengthening, so the chains can become extremely long. The P + Epsilon attack is a method of exploiting the dominant strategies of the participants in the network. PoW-based blockchains are usually vulnerable to this type of attack [55]. When attackers give participants a pay-out in order to gain an advantage, a payoff matrix is used where the dominant tactic facilitates the achievement of the attacker's objectives. As a result of the attack, the participants do not receive any compensation, and the attacker receives the entire amount. This key statistical finding is based on an ad hoc selection model.

2.5. Hybrid Approaches Related to 51% Attacks

The term 51% attacks refers to situations in which an attacker has 51% of the hashing power. As part of this attack, a private blockchain is created and completely disconnected from the actual chain edition. It is later introduced to the network as a real chain, which allows for a double-spending attack [47] Additionally, given that blockchain policy follows the most extended chain rule [56], if attackers gain 51% or more of the threat, they will push the longest chain by convincing network nodes to obey their chain. However, 51% of computational power is not strictly sufficient, so double spending is still possible if an attacker has less than half of the computational power [48]. The odds of success are low. A blockchain attack becomes increasingly expensive when the entire network acquires increased hash power. A cryptocurrency with a high network hash rate may also be resilient to 51% attacks. To overcome 51% attacks, several studies and developments are being conducted. Researchers have proposed mixing proof mechanisms to eliminate 51% attacks. For PoW to be applied to a working network, the attacker must gain more than 50% of the processing capacity and more than 50% of the network wealth. This task is highly challenging for a user. In addition, the total cost should be considerably lower than the profit that an attacker might earn. The attacker's costs are much higher than the benefit in this form of a hybrid network.

Additionally, this hybridisation implements other security measures to counter the attack [28]. Different studies and innovations have recommended different prevention methods, but they have several limitations in common. Komodo [57] introduced dPOW consensus, which takes a snapshot of the blockchain every 10 min and stores it in the blockchain. One of the more recent developments implemented by Horizen at Zen Coin is to delay the block in order to slow down the creation of blocks [29]. Casper and Decred provided a second hybrid PoS consensus using a BFT model with a two-thirds vote mechanism in the first 50 networks. This voting process is independent, resulting in unintended delays and an inconsistent block interval [30]. An alternative hybrid consensus algorithm was proposed by the authors in [31], namely, fork-free hybrid consensus with versatile proof-of-activity and the hybrid PoW-PoS-PoA algorithm. The authors introduced a technique where all PoW chains are created simultaneously and submitted to a committee for review. The committee determines and approves the most robust chain as the main chain. A weighted calculation amongst the committee members determines which chain is the best. On the basis of PoW power and PoS capacity, the weight of each committee member is calculated. This algorithm can also mitigate 51% attacks. However, other issues may affect the blockchain. One of the primary issues is determining how to distribute newly produced block rewards [27]. Another issue is that the interval between block generation is often inconclusive, which is directly related to the recently created currency [49]. Table 2 summarises various hybrid approaches and other solutions discussed in the literature.

Table 2. Comparison of the proposed hybrid mechanism and other hybrid solutions.

Solution	P2P Protocol	Proportional Gain Proportion	Fair Voting Mechanism	Stable Block Time	Autonomous Network	Punishment for Malicious Nodes
Komodo [57]	✓	✗	✓	✓	✓	✗
ZEN [29]	✗	✗	✓	✓	✗	✓
Decred [30]	✗	✗	✗	✗	✗	✗
Casper [30]	✗	✗	✗	✗	✗	✓
Proposed mechanism	✓	✓	✓	✓	✓	✓

In conclusion, this study proposes a hybrid algorithm that combines PoS and PoW mechanisms to provide a fair mining reward to both the miner and validator. By maintaining a precise block generation interval with difficulty adjustment in power mining and a likelihood measurement according to the stake's mature stake balance, the system not only resolves the 51% attack but also provides stakeholders and investors with a uniform distribution of mining rewards.

3. Proposed Hybrid Approach

In this study, we combine two consensus mechanisms for a fair mining reward for the miner and validator into a hybrid model. Assume that the behaviour of nodes is likely to be known as the most massive chain. As a result, the first block generated in this model is usually referred to as the main chain, along with the majority of the nodes in the network. In Figure 4, we present our proposed finite state automata (FSA) model for the block-forging process.

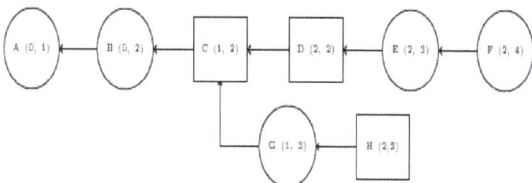

Figure 4. FSA for the block-forging process.

In Figure 4, the square blocks correspond to PoW, and the circles correspond to PoS. The arrows represent canonical chains. Under certain conditions, PoW and PoS blocks are mined and staked in random order, and the possibility of reaching a consensus is approximately 50% [41]. Nx is the set of all positive integers smaller than 2^x, and block $b = (f_p, f_{sr}, f_{tr}, f_d, f_{ts}, f_{tx})$, where $f_p, f_{sr}, f_{tr}, f_d \in N256$, $f_d \in N64$ and f_{tx} is a linked list. Table 3 provides a description of these elements.

Table 3. Description of the elements of the proposed algorithm.

Element	Description
f_p	Parent block hash
f_{sr}	Root node hash tree after all transactions are executed
f_{tr}	Root node hash for every transaction
f_d	Computing (mining/stacking) complexity
f_{tx}	Transaction included inside a block
f_{ts}	Time of generated blocks

Miners working in a conventional PoW mining setup require a step-by-step implementation, as depicted in Algorithm 1. With the mining difficulty parameter d_w and the 256-bit-long function $hash(\cdot)$, miners can solve complex problems within this rule, as shown in Equation (1).

$$hit = hash(b) \leq 2256/dw \quad (1)$$

After completing mining, several rewards are provided, and their mining power is proportional to the computation power.

Algorithm 1. Mining for the PoW mechanism

1	**Procedure** MINING PoW (δ)
2	$k \leftarrow$ GetBestChain
3	$z_1 \leftarrow$ GetLastBlock(k)
4	$z_2 \leftarrow$ GetSecondLastBlock(k)
5	$diff \leftarrow$ GetComplexity(z_1, z_2)
6	$trxs \leftarrow$ GetMemoryPoolTrxs()
7	$z \leftarrow$ CreateBlockTemplate(k,trxs)
8	do
9	$thesolution \leftarrow$ ProofofWork(z)
10	**while** $thesolution > 2^{256}/diffs$
11	$z \leftarrow$ Finalize(z,thesolution)
12	Import & Propagate(z)
13	end

Figure 5 illustrates the proposed hybrid model flow, in which the mining process begins with the identification of stake parameters. These are mature balancing parameters for stakes, coinage, the synchronisation of timestamps, the weights of individual nodes and the weight of the entire network. After the initial validations and time sync prerequisite tests, we add the PoW nonce discovery loop. Next, an empty block template is created. The PoW loop then locates a valid nonce to generate a valid hash. The block contains individual transactions that cannot be arbitrarily modified. Other block records, such as timestamps and earlier hash blocks, are irreversible. Therefore, to adjust the hash and achieve a correct pattern, the nodes use the nonce arbitrary field. As part of the PoW loop, miners start with 0 and continue to increase the nonce and produce hashes whilst merging this nonce with other block data. When a correct hash that meets the requirements of the block hash is discovered, the peer achieves success in mining. A complexity factor is added for the block interval to be preserved.

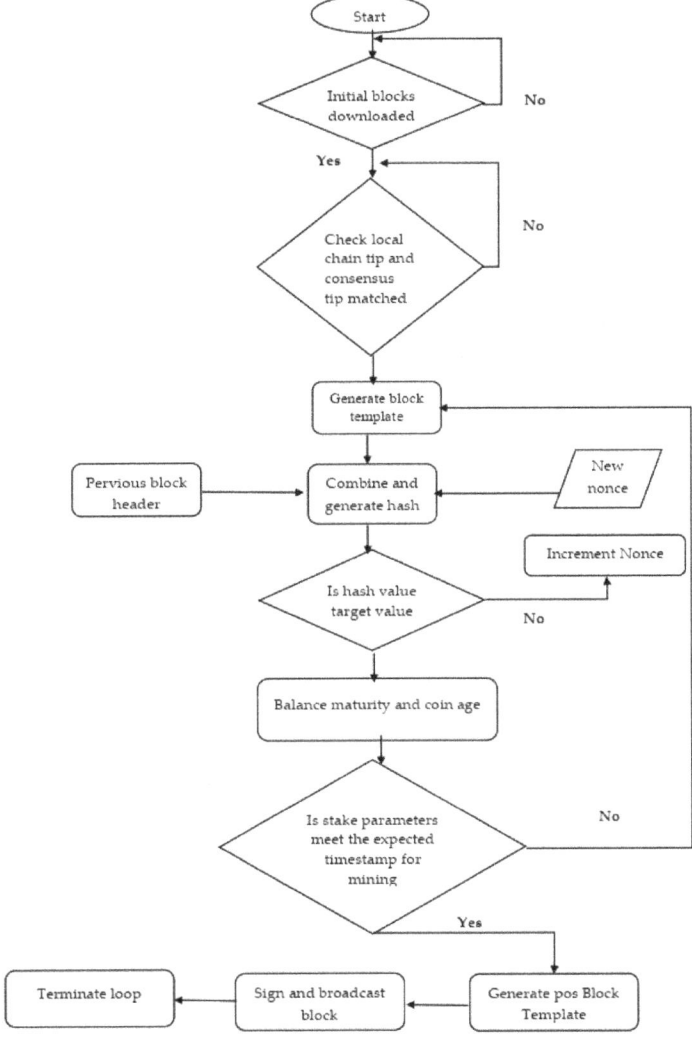

Figure 5. Process of the proposed hybrid model with supported features.

By applying this approach, we can achieve excellent control over the generation time interval for blocks. In addition, the benefits of mining and transaction fees are evenly distributed amongst investors. Apart from this mining method, another security measure is implemented to secure the network against the misbehaviour of nodes by preventing these nodes in a predefined period. A minimum of one hour is required to ban the simulation setting. Whenever a peer node obstructs, the peer is banned for one hour from the network.

The protocol imposes an additional restriction that all nodes must be fairly validated. Both network nodes are equally weighted with regard to decision making. It involves validating a rich node and judging a weak node fairly. The two peer nodes share the same code and weight. Given that the mining process involves a degree of risk, every node can verify transactions and blocks. The frequency of the chance depends on its staking capacity.

Furthermore, our proposed method incorporates PoS and PoW into a stochastic coherence process without sacrificing availability, and a decentralised stack is essential. Considering that the systems run based on computations and stakes in the network, we define a rule-based forking mechanism to ensure that new blocks are produced between the two consensus types. By examining how much effort is made and the rewards obtained by stacker and miner devices, which should be fair, this study demonstrates its novelty. This simulation proposes a minor tweak to the difficulty adjustment, as indicated in Equation (2).

$$td_s, c_0 = \text{argmax } td_{wi} \cdot td_{si}; i \in \{1,...,N\} \quad (2)$$

In general, the algorithm chooses the appropriate complexity to match the inside network's hash/stake power. However, this choice is sometimes gradual. Consider, for example, that stake complexity is approximately 10 times greater than miner complexity. There is a 10x increase in stake in comparison with its hash rate. Unlike PoW, each stacker processes several numbers and keys.

$$seed_{t+1} = \text{sign}(seed_t, sk) \quad (3)$$

When this condition is met, a stake block can be produced.

$$\ln(hash(seed)/2^{256}) \mid \cdot d_s \leq V \cdot \Delta, \quad (4)$$

where V is the amount of the computation unit and Δ is the time from the last block. To define the algorithm target, we should have t as the target time and $2t$ becomes the target time for PoS and PoW. Double-spending attacks take place when an individual has more than 51% of the peer network either as a miner or as a stacker. The dominant attacker is assumed to have power defined with a and b notations, and the ordinary nodes are defined with c and d notations. The hash (PoW) block generation rate is $\lambda_w = \frac{w}{d_\omega}$, where (w) is the hash rate. During the simulation, the number of blocks were generated using random variable $X \sim PoS(\lambda_w)$, and $E(X) = \lambda_w$. For example, assume that Y_w is a notation of the total difficulty of the mining process; thus, $E(Y_w) = E(X) \cdot d_w$.

$$E(Y_\omega) = \frac{w}{d_\omega} * d_\omega = w \quad (5)$$

Similarly in Algorithm 2, the PoS block generation rate is declared as $\lambda_w = \frac{s}{d_s}$, where s is the amount of stake.

The notation Y_s is the total stack difficulty.

$$E(Y_s) = \frac{w}{d_s} * d_s = s \quad (6)$$

Meanwhile, the attacker's chain contains a weight within the expected period.

$$(td_{wc} + a \cdot t) \cdot (td_{sc} + b \cdot t) \quad (7)$$

The ordinary nodes' rules are defined as follows:

$$(td_{wc} + c \cdot t) \cdot (td_{sc} + d \cdot t), \quad (8)$$

where td_w and td_s represent the total difficulty/complexity of mining and stacking blocks, respectively. According to the prospectus of the attacker, overtaking another chain requires the attacker to possess greater power than the normal nodes, which results in network inequality.

$$td_{sc} \cdot (a - c) + td_{wc} \cdot (b - d) + (ab - cd) \cdot t \geq 0. \tag{9}$$

Algorithm 2. Stacking Algorithm

1   **Procedure** STAKEBLOCK(δ,pk,sk)
2    $k \leftarrow$ GetBestNode
3    $z_1 \leftarrow$ GetLastBlock(k)
4    $z_2 \leftarrow$ GetSecondLastBlock(k)
5   stakes $\leftarrow$ GetPoStake(k,pk)
6   diffs $\leftarrow$ GetComplexity(z_1,z_2)
7   tms $\leftarrow$ GetTimestamp(z_1)
8   seeds $\leftarrow$ GetSeed(z_1)
9   seeds $\leftarrow$ Sign(seeds,sks)
10    $\Delta \leftarrow$ diffs $\cdot$ ln(hash(seeds)/2^{256})/stake
11   **Do**
12   sleep(1)
13   **While** $\varphi <$ tms $+ \Delta$
14   trxs $\leftarrow$ GetMemoryPoolTrxs()
15   z $\leftarrow$ CreateBlockTemplate(k,trxs,seeds)
16   z $\leftarrow$ Final(z,sk)
17   Import & Propagate(z)

Double Spending Attack Prevention Scenario

Only when both a PoW block and a PoS block confirm a transaction should it be considered confirmed on the blockchain. A transaction should not be considered confirmed when only PoS blocks confirm it because PoS blocks can be minted over multiple conflicting chains. As long as people refrain from erroneously considering 1-PoS-confirmed transactions as confirmed, this should not be an issue.

Furthermore, a transaction should not be considered confirmed when only PoW blocks confirm it because this could lead to double spending by an attacker using a 51% attack. This attack is much harder than double spending for someone accepting only PoS blocks as confirmation, but it is likely to be much easier than it is for today's Bitcoin because the new algorithm reduces the cost of mining (which in turn reduces the system's hash power by nature). For this reason, both PoW and PoS should be used to confirm or finalise transactions.

The expenditures required to launch a 51% attack are much greater than those for PoW for a given amount of honest mining. Hence, an attacker requires an amount of hash power equal to the honest hash power (which in an equilibrium case results in the attacker possessing 100% of the hash power). In addition, an attacker needs to own a considerable amount of hash stake. Given that the longest chain is determined by multiplying PoW and PoS accumulated difficulties, even if a single miner accumulates 90% of the mining power, it would not be able to produce a significantly longer chain without also owning more than 11% of current coins in circulation.

Considering a scenario in which the attacker attempts to create an additional sidechain and reveals it at a, we assume that the attacker has a hash power and stake power of (a,b), and the fair nodes have (c,d). Let Y_w be the total mining difficulty. Then, $E(Y_w) = E(X) * d_w$. This has been given in Equation (1), where the PoS block generation rate $\lambda_w = \frac{n}{d_2}$, where s is the stake and Y_s is the total mining difficulty presented in Equation (6). The total mining difficulty is an integration of the hash rate over time and vice versa of the stake over time. In duration t, the malicious chain has an expected weight of $(td_{wc} + a \cdot t) \cdot (td_{sc} + b \cdot t)$, and the fair nodes' chain has $(td_{wc} + c \cdot t) \cdot (td_{sc} + d \cdot t)$, where td_w and td_s are the total difficulty for PoW and PoS from the genesis block, respectively.

For the attacker to gain the fair nodes' chain, the malicious nodes need to have a longer chain than the fair nodes' chain, which further leads to the following inequality: $ld_{sc} \cdot (a - c) + ld_{wc} \cdot (b - d) + (ab - cd) \cdot l \geq 0$. Given that this attack can only occur if the creation of blocks is free, we assume that the attacker will attempt to attack by using only PoS blocks. Assume that

$$td_a = \sum_{i=1..II_{w-n}} d_{wi} \cdot \sum_{j=1...I_s} d_{sj} = \left((H_w - n) \cdot \overline{d_w}\right) \cdot \left(H_s \cdot \overline{d_s}\right), \tag{10}$$

where td_a indicates the total complexity of the malicious chain. Even if the attacker holds the entire active stake and the total voting power remains unchanged, the best-case scenario is an identical t_d for the main chain. The projected maximum number of blocks that the LRA can create is $(\phi - t_{N_w-n})/2t$ because the protocol forbids the creation of new blocks. If an attacker can increase his stake power through block rewards, then his chances of success increase with time. Specifically, the assailant must reach

$$\left(H_s \cdot \left(\overline{d_s} + \Omega\right)\right) > \left(H_w \cdot \overline{d_w}\right) \cdot \left(H_s \cdot \overline{d_s}\right) \Omega > \frac{H_w \cdot \overline{d_w}}{H_s}. \tag{11}$$

Assuming that the primary chain's forging power is static (i.e., not subject to change), $N_w = N_s$ is modified to reflect the extra power an attacker would require to equal the main chain's strength (expressed in difficulty). It must be more challenging than the PoW chain itself. Further research is required to determine how long it takes an attacker to gain access to increased difficulty, but the premise is that this process of gaining power gradually via block rewards occurs over a long period.

4. Experimental Results

During the implementation phase, we set the simulation in such a way that the hash output is uniformly distributed between miners and stakes. The difficulty was adjusted to $\alpha = 0.01$, the stacker power was set to S = [80, 40, 20, 15, 10, 5, 5, 5, 5, 5] and the miner power was set to M = [32, 16, 8, 6, 4, 2, 2, 2, 2, 2]. Numbers have already been set to show the linearity of the exponential rise in computational power. To begin with, the block time was set to 20 s in t, with a duration of 90 days per entire chain.

During the simulation, a total of 385,479 blocks were generated, out of which 192,688 were stake blocks and 192,791 were mining blocks. As shown in Figure 6, the rewards were proportional to computing power (stake/mining), which was considered a fair outcome. The target block time was 20 s, resulting in a stake block time of 40 s and a mining block time of 40 s with an average rate of $Ps \in S$ s/ds, $Pm \in M$ m/dm and $Ps \in S$ s/ds + $Pm \in M$ m/dm.

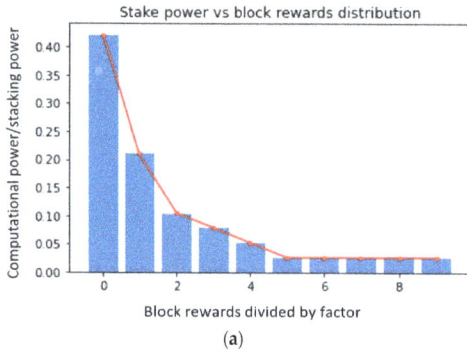
(a)

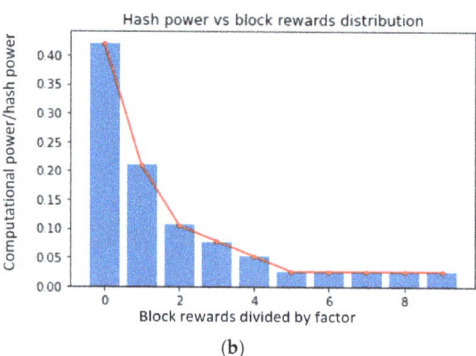
(b)

Figure 6. Experimental results of (**a**) stake power vs. block rewards distribution and (**b**) hash power vs. block rewards distribution.

During the experiment, which we ran on the Google Colab platform, we determined that the initial simulation would consume not more than 200 MB of RAM, as shown in Table 4.

Table 4. Machine computational power.

	Free RAM	Process Size
Beginning	12.8 GB	118.9 MB
End	12.7 GB	383.7 MB

Figure 6a,b illustrate the stake and hash power results over the block rewards distribution. The mean and standard deviation of time are presented in Table 5.

Table 5. Mean and standard deviation of time.

Parameters	Mean	Standard Deviation
All blocks	20.217	19.977
PoS blocks	40.464	40.325
PoW blocks	40.406	40.251

According to the simulation results, the attacker side that dominates the network with more than 51% computation power cannot easily launch the attack because the fork mechanism has a split rule between PoW and PoS implementations. Hence, to take over this network and launch an attack, the attacker needs to command over 100% of the system, which is impossible on a consensus blockchain node. Figure 7 shows that the chain power is demonstrated over the block time generation and distribution, and Figure 8 shows the proposed hybrid model computational power over PoW and PoS block time distributions.

We began the simulation experiment and based on previous data, we set the computational power for mining to 76, which is the same block size. The results are shown in Table 6.

Additionally, we discovered that the combined PoW and PoS protocol limits the effective computational power to 52.31. The results of the miner computational power needed for the hybrid mechanism are presented in Table 7.

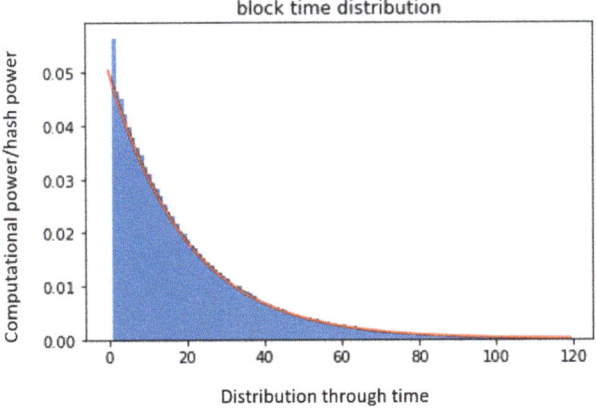

Figure 7. Power vs. block time distribution.

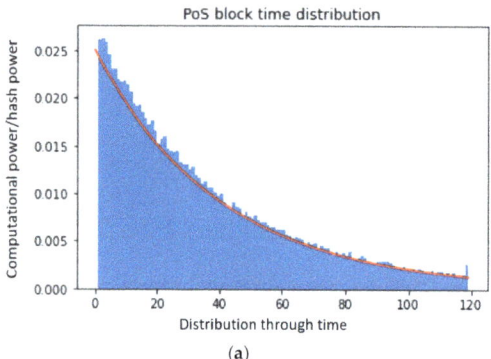
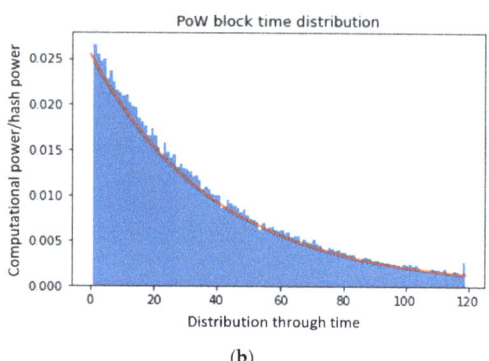

Figure 8. Experimental results of (**a**) computational power vs. PoS block time distribution and (**b**) computational power vs. PoW block time distribution.

Table 6. Miners' computational power before implementing the proposed hybrid mechanism.

Miners	Computational Power Percentage
Miner 0	23%
Miner 1	21.24%
Miner 2	17.36%
Miner 3	12.87%
Miner 4	8.56%
Miner 5	4.2%
Miner 6	4.3%
Miner 7	4.26%
Miner 8	4.2%

Table 7. Miners' computational power with the proposed hybrid mechanism (PoW and PoS).

Miners	Computational Power Percentage
Miner 0	15.19%
Miner 1	11.18%
Miner 2	7.94%
Miner 3	6%
Miner 4	4%
Miner 5	2%
Miner 6	2%
Miner 7	2%
Miner 8	2%
Total	52.31

Mining power and actively minting coins may also be used to calculate the attack's cost. They may be used as a rough estimate of the cost of mining power because they are closely related to miner earnings (fees and coin base incentives). The active stake can be determined because the amount of Satoshi released every second is inversely proportional to the stake difficulty. By dividing the amount of Satoshi issued into equal parts for each PoS block, we can estimate the total amount of Satoshi currently being mined. To compute the income per block required to sustain the attack cost, these measures may be used to determine an attack–cost objective (e.g., a particular number of Bitcoins or a certain percentage of the total number of Bitcoins mined to date). In turn, this information can be used to dynamically alter the block size and ensure that the block income continues to support the set attack cost goal. Therefore, mining earnings will be increasingly predictable, a certain degree of security will be maintained, and costs will be reduced.

The majority of honest minters make the mistake of making coins on a chain that they believe will last the longest, only to have their efforts thwarted by a competitor's longer chain. This situation means that several law-abiding minters are penalised for minting. If the fine does not exceed the revenue from minting one block, the expected revenue from the effort to mint should exceed zero. A small fee is likely to have a substantial impact, so the projected revenue from minting should be equal to the overall revenue from minting. Ultimately, this depends on whether dishonest minting on a short chain benefits the dishonest minter. Therefore, how much of a penalty should be imposed is debatable.

Given that PoS blocks have little influence on whether an attacker will be successful in performing an orphan-based mining monopoly attack unless the attacker controls a substantial portion of the coins actively being mined, punishing minters who minted over another PoS block would double the collateral damage. Consequently, minter punishment proofs will be invalid if the most recent PoW block is shared by the minted block and the current block. Further research is needed to determine how much stake an attacker needs to possess in order to perform a mining monopoly attack effectively.

If a PoW block has more than one option (e.g., a collision leaving one orphaned), minters may refuse to mint so as to avoid the penalty. If they do so, they will miss out on the most probable benefit of their actions (which would be much greater). Therefore, the likelihood of this behaviour occurring is very low because the predicted benefits exceed 0 by a factor of two.

5. Conclusions

Bitcoin's popularity and success are primarily related to the underlying blockchain technology, which is a genuinely unchanging and highly protected distributed ledger governed by a peer-to-peer consensus. This study conducted a comprehensive analysis to build the hybrid cryptocurrency PoW-PoS, which can resolve the 51% attack in the most feasible and advanced way possible. The proposed hybrid model can prevent the attack by mixing PoW and PoS in one thread with a strict time spacing for block generation to achieve a resilient and robust agreement amongst P2P network nodes and guarantee a benefit distribution that is in line with stakeholders' investment ratios. The results showed that we successfully implemented a hybrid consensus protocol for blockchain that combines mining and stacking. The hybrid protocol creates a fair mechanism for miners and stakers. Furthermore, each block's period is added to provide a double-spending function for every distribution even though the attacker has more than 51% control over the network. We examined the shortcomings of consensus protocols and security techniques to reveal their main weaknesses. A hybrid model that combines PoW and PoS was then successfully implemented. The system incorporates hardware and economic security without compromising availability, predictability or decentralisation. According to the empirical evidence provided in results Section 4, the proposed protocol is fair and scalable to an arbitrary number of miners and stakes.

In the future, we will conduct a more comprehensive analysis of network stability with additional types of miners and stackers and another simulation based on game theory. A highly economical approach for all mechanisms will also be explored in the future.

Author Contributions: Conceptualization, N.A.A. and A.M.; methodology, N.A.A. and A.M.; software, N.A.A.; validation and formal analysis, N.A.A., A.M. and S.M.F.; investigation, A.M. and S.M.F.; writing—original draft preparation, N.A.A. and A.M.; writing—review and editing, S.M.F. and N.E.; visualization, N.A.A. and A.M.; supervision, S.M.F.; project administration, A.M.; Funding N.E. All authors have read and agreed to the published version of the manuscript.

Funding: The authors would like to acknowledge the support of Prince Sultan University for paying the article processing charges (APC) of this publication.

Institutional Review Board Statement: Not applicable.

Informed Consent Statement: Not applicable.

Data Availability Statement: Not applicable.

Conflicts of Interest: The authors declare no conflict of interest.

References

1. Herbaut, N.; Negru, N. A Model for Collaborative Blockchain-Based Video Delivery Relying on Advanced Network Services Chains. *IEEE Commun. Mag.* **2017**, *55*, 70–76. [CrossRef]
2. Kang, J.; Yu, R.; Huang, X.; Maharjan, S.; Zhang, Y.; Hossain, E. Enabling Localized Peer-to-Peer Electricity Trading Among Plug-in Hybrid Electric Vehicles Using Consortium Blockchains. *IEEE Trans. Ind. Inform.* **2017**, *13*, 3154–3164. [CrossRef]
3. Dwivedi, A.D.; Srivastava, G.; Dhar, S.; Singh, R. A decentralized privacy-preserving healthcare Blockchain for IoT. *Sensors* **2019**, *19*, 326. [CrossRef] [PubMed]
4. Karamitsos, I.; Papadaki, M.; Al Barghuthi, N.B. Design of the Blockchain smart contract: A use case for real estate. *J. Inf. Secur.* **2018**, *9*, 177–190. [CrossRef]
5. Li, M.; Shen, L.; Huang, G.Q. Blockchain-enabled workflow operating system for logistics resources sharing in E-commerce logistics real estate service. *Comput. Ind. Eng.* **2019**, *135*, 950–969. [CrossRef]
6. Liu, Z.; Li, Z. A Blockchain-based framework of cross-border e-commerce supply chain. *Int. J. Inf. Manag.* **2020**, *52*, 102059. [CrossRef]
7. Wu, B.; Li, Y. Design of evaluation system for digital education operational skill competition based on Blockchain. In Proceedings of the 2018 IEEE 15th International Conference on e-Business Engineering (ICEBE), Xi'an, China, 12–14 October 2018; IEEE: New Piscataway, NJ, USA, 2018; pp. 102–109.
8. Zhang, S.; Rong, J.; Wang, B. A privacy protection scheme of smart meter for decentralized smart home environment based on consortium Blockchain. *Int. J. Electr. Power Energy Syst.* **2020**, *121*, 106140. [CrossRef]
9. Lee, Y.; Rathore, S.; Park, J.H.; Park, J.H. A Blockchain-based smart home gateway architecture for preventing data forgery. *Hum.-Cent. Comput. Inf. Sci.* **2020**, *10*, 1–14. [CrossRef]
10. Makhdoom, I.; Zhou, I.; Abolhasan, M.; Lipman, J.; Ni, W. PrivySharing: A Blockchain-based framework for privacy-preserving and secure data sharing in smart cities. *Comput. Secur.* **2020**, *88*, 101653. [CrossRef]
11. Kumar, R.R.; Menon, S.; Nair, N.S. Blockchain Solutions for Security Threats in Smart Industries. In Proceedings of the 2020 Fourth International Conference on Computing Methodologies and Communication (ICCMC), Erode, India, 11–13 March 2020; IEEE: Piscataway, NJ, USA, 2020; pp. 756–763.
12. Rathee, G.; Garg, S.; Kaddoum, G.; Choi, B.J. A decision-making model for securing IoT devices in smart industries. *IEEE Trans. Ind. Inform.* **2020**, *17*, 4270–4278. [CrossRef]
13. Miller, A.; Juels, A.; Shi, E.; Parno, B.; Katz, J. Permacoin: Repurposing bitcoin work for data preservation. In Proceedings of the 2014 IEEE Symposium on Security and Privacy, Berkeley, CA, USA, 18–24 May 2014; IEEE: Piscataway, NJ, USA, 2014; pp. 475–490.
14. Memon, R.A.; Li, J.P.; Ahmed, J. Simulation model for Blockchain systems using queuing theory. *Electronics* **2019**, *8*, 234. [CrossRef]
15. Christidis, K.; Devetsikiotis, M. Blockchains and smart contracts for the internet of things. *IEEE Access* **2016**, *4*, 2292–2303. [CrossRef]
16. Akbar, N.A.; Sunyoto, A.; Arief, M.R.; Cesarendra, W. Reducing overhead of self-stabilizing byzantine agreement protocols for blockchain using http/3 protocol: A perspective view. *Sinergi* **2021**, *25*, 381. [CrossRef]
17. Xiang, H.; Ren, Z.; Zhou, Z.; Wang, N.; Jin, H. AlphaBlock: An Evaluation Framework for Blockchain Consensus Protocols. *arXiv* **2020**, arXiv:2007.13289.
18. Mingxiao, D.; Xiaofeng, M.; Zhe, Z.; Xiangwei, W.; Qijun, C. A review on consensus algorithm of Blockchain. In Proceedings of the 2017 IEEE International Conference on Systems, Man, and Cybernetics (SMC), Banff, AB, Canada, 5–8 October 2017; IEEE: Piscataway, NJ, USA, 2017; pp. 2567–2572.
19. Antonopoulos, A.M. *Mastering Bitcoin: Unlocking Digital Cryptocurrencies*; O'Reilly Media, Inc.: Sebastopol, CA, USA, 2014.
20. Wang, W.; Hoang, D.T.; Hu, P.; Xiong, Z.; Niyato, D.; Wang, P.; Wen, Y.; Kim, D.I. A Survey on Consensus Mechanisms and Mining Strategy Management in Blockchain Networks. *IEEE Access* **2019**, *7*, 22328–22370. [CrossRef]
21. Sharkey, S. Alt-PoW: An Alternative Proof-of-Work Mechanism. 2018. Available online: https://www.researchgate.net/publication/328150068 (accessed on 1 October 2021).
22. Pilkington, M. Blockchain technology: Principles and applications. In *Research Handbook on Digital Transformations*; Edward Elgar Publishing: Cheltenham, UK, 2016.
23. Dannen, C. *Introducing Ethereum and Solidity*; Apress: Berkeley, CA, USA, 2017; Volume 1.
24. Ren, W.; Hu, J.; Zhu, T.; Ren, Y.; Choo, K.K.R. A flexible method to defend against computationally resourceful miners in Blockchain proof of work. *Inf. Sci.* **2020**, *507*, 161–171. [CrossRef]
25. Shanaev, S.; Shuraeva, A.; Vasenin, M.; Kuznetsov, M. Cryptocurrency value and 51% attacks: Evidence from event studies. *J. Altern. Invest.* **2019**, *22*, 65–77. [CrossRef]
26. Sayeed, S.; Marco-Gisbert, H. Assessing Blockchain consensus and security mechanisms against the 51% attack. *Appl. Sci.* **2019**, *9*, 1788. [CrossRef]
27. Andoni, M.; Robu, V.; Flynn, D.; Abram, S.; Geach, D.; Jenkins, D.; McCallum, P.; Peacock, A. Blockchain technology in the energy sector: A systematic review of challenges and opportunities. *Renew. Sustain. Energy Rev.* **2019**, *100*, 143–174. [CrossRef]

28. Ghosh, A.; Gupta, S.; Dua, A.; Kumar, N. Security of Cryptocurrencies in Blockchain technology: State-of-art, challenges and future prospects. *J. Netw. Comput. Appl.* **2020**, *163*, 102635. [CrossRef]
29. Kishor Datta Gupta, A.R. A Hybrid POW-POS Implementation Against 51% Attack in Cryptocurrency System. Available online: https://www.researchgate.net/publication/337831342 (accessed on 13 October 2020).
30. Burkhard Stiller, M.F. *Communication Systems XII*; Department of Informatics (IFI), University of Zurich: Zurich, Switzerland, 2019.
31. Liu, Z.; Tang, S.; Chow, S.S.; Liu, Z.; Long, Y. Fork-free hybrid consensus with flexible proof-of-activity. *Future Gener. Comput. Syst.* **2019**, *96*, 515–524. [CrossRef]
32. Monrat, A.A.; Schelén, O.; Andersson, K. A Survey of Blockchain from the Perspectives of Applications, Challenges and Opportunities. *IEEE Access* **2019**, *7*, 117134–117151. [CrossRef]
33. Atzori, M. Blockchain Technology and Decentralized Governance: Is the State Still Necessary? *J. Gov. Regul.* **2017**, *6*, 45–62. [CrossRef]
34. Rui Zhang, R.X. Security and Privacy on Blockchain. *Acm Comput. Surv.* **2019**, *52*, 1–34. [CrossRef]
35. Zhou, Q.; Huang, H.; Zheng, Z.; Bian, J. Solutions to scalability of Blockchain: A survey. *IEEE Access* **2020**, *8*, 16440–16455. [CrossRef]
36. Xie, J.; Yu, F.R.; Huang, T.; Xie, R.; Liu, J.; Liu, Y. A survey on the scalability of Blockchain systems. *IEEE Netw.* **2019**, *33*, 166–173. [CrossRef]
37. Hertig, A. Cat Fight? Ethereum Users Clash Over CryptoKitties. 7 December 2017. Available online: https://www.coindesk.com/markets/2017/12/07/cat-fight-ethereum-users-clash-over-cryptokitties/ (accessed on 1 October 2021).
38. Niranjanamurthy, M.; Nithya, B.N.; Jagannatha, S. Analysis of Blockchain technology: Pros, cons and SWOT. *Clust. Comput.* **2019**, *22*, 14743–14757. [CrossRef]
39. Baliga, A. Understanding Blockchain Consensus Model. 2017. Available online: https://pdfs.semanticscholar.org/da8a/37b10bc1521a4d3de925d7ebc44bb606d740.pdf (accessed on 1 October 2021).
40. Solving the Byzantine Generals Problem with Delegated Proof of Stake (DPoS). 2018. Available online: https://www.radixdlt.com/post/what-is-delegated-proof-of-stake-dpos (accessed on 27 September 2020).
41. Gramoli, V. From Blockchain Consensus Back to Byzantine Consensus. Data61-CSIRO and University of Sydney Australia. 1 September 2018. Available online: https://www.researchgate.net/publication/319984012 (accessed on 1 October 2021).
42. Gutteridge, D. Japanese Cryptocurrency Monacoin Hit by Selfish Mining Attack. Available online: https://www.ccn.com/japanese-cryptocurrencymonacoin-hit-by-selfish-mining-attack/ (accessed on 1 October 2021).
43. Redman, J. Bitcoin Gold 51% Attack. Available online: https://news.bitcoin.com/bitcoingold-51-attackednetwork-loses-70000-in-double-spends/ (accessed on 1 October 2021).
44. Tassev, L. Bitcoin in Brief Monday: Zencash Targeted in 51 Hijacked for Ransom. 2018. Available online: https://news.bitcoin.com/bitcoin-in-briefmonday-zencash-targeted-in-51-attackticketfly-hijackedfor-ransom/ (accessed on 1 October 2021).
45. Chauhan, A.; Malviya, O.P.; Verma, M.; Mor, T.S. Blockchain and scalability. In Proceedings of the 2018 IEEE International Conference on Software Quality, Reliability and Security Companion (QRS-C), Lisbon, Portugal, 19–20 July 2018; IEEE: Pisataway, NJ, USA, 2018; pp. 122–128.
46. Xu, X.; Weber, I.; Staples, M.; Zhu, L.; Bosch, J.; Bass, L.; Pautasso, C.; Rimba, P. A taxonomy of Blockchain-based systems for architecture design. In Proceedings of the 2017 IEEE International Conference on Software Architecture (ICSA), Gothenburg, Sweden, 3–7 April 2017; IEEE: Piscataway, NJ, USA, 2017; pp. 243–252.
47. Natoli, C.; Gramoli, V. The balance attack against proof-of-work Blockchains: The R3 testbed as an example. *arXiv* **2016**, arXiv:1612.09426.
48. Natoli, K. Cryptoeconomics: Paving the Future of Blockchain Technology. 2017. Available online: https://hackernoon.com/cryptoeconomics-paving-the-future-of-Blockchain-technology-13b04dab97 (accessed on 10 October 2020).
49. Gervais, A.; Karame, G.O.; Wüst, K.; Glykantzis, V.; Ritzdorf, H.; Capkun, S. On the security and performance of proof of work Blockchains. In Proceedings of the 2016 ACM SIGSAC Conference on Computer and Communications Security, Vienna, Austria, 24–28 October 2016; Association for Computing Machinery: New York, NY, USA, 2016; pp. 3–16.
50. Kaushik, K.; Dahiya, S.; Singh, R.; Dwivedi, A.D. Role of Blockchain in Forestalling Pandemics. In Proceedings of the 2020 IEEE 17th International Conference on Mobile Ad Hoc and Sensor Systems (MASS), Delhi, India, 10–13 December 2020; IEEE: Pisataway, NJ, USA, 2020; pp. 32–37.
51. Thin, W.Y.M.M.; Dong, N.; Bai, G.; Dong, J.S. Formal analysis of a proof-of-stake Blockchain. In Proceedings of the 2018 23rd International Conference on Engineering of Complex Computer Systems (ICECCS), Melbourne, Australia, 12–14 December 2018; IEEE: Piscataway, NJ, USA, 2018; pp. 197–200.
52. Asolo, B. Delegated Proof of Stake (DPOS) Explained. 2018. Available online: https://www.mycryptopedia.com/delegated-proof-stake-dpos-explained/ (accessed on 1 July 2018).
53. Sharma, A. Understanding Proof of Stake Through Its Flaws. Part 3 Long Range Attacks. 2018. Available online: https://medium.com/@abhisharm/understanding-proof-of-stake-through-its-flaws-part-3-longrange-attacks-672a3d413501 (accessed on 2 October 2020).
54. Buterin, V. Long-Range Attacks: The Serious Problem with Adaptive Proof of Work. 2014. Available online: https://blog.ethereum.org/2014/05/15/long-range-attacks-the-serious-problem-with-adaptiveproof-of-work/ (accessed on 2 October 2020).

55. Buterin, V. The P + Epsilon Attack. 2015. Available online: https://blog.ethereum.org/2015/01/28/pepsilon-attack/ (accessed on 4 October 2020).
56. Vitalik Buterin. Selfish Mining: A 25% Attack against the Bitcoin Network. 2013. Available online: https://bitcoinmagazine.com/articles/selfish-mining-a-25-attack-against-the-bitcoin-network-1383578440/ (accessed on 14 October 2020).
57. ChainZilla. Solutions to 51% Attacks and Double Spending. Medium. 2020. Available online: https://medium.com/chainzilla/solutions-to-51-attacks-and-double-spending-71526be4bb86 (accessed on 13 October 2020).

 *future internet*

Article

EconLedger: A Proof-of-ENF Consensus Based Lightweight Distributed Ledger for IoVT Networks

Ronghua Xu, Deeraj Nagothu and Yu Chen *

Department of Electrical and Computer Engineering, Binghamton University, SUNY, Binghamton, NY 13905, USA; rxu22@binghamton.edu (R.X.); dnagoth1@binghamton.edu (D.N.)
* Correspondence: ychen@binghamton.edu; Tel.: +1-607-777-6133

Abstract: The rapid advancement in artificial intelligence (AI) and wide deployment of Internet of Video Things (IoVT) enable situation awareness (SAW). The robustness and security of IoVT systems are essential for a sustainable urban environment. While blockchain technology has shown great potential in enabling trust-free and decentralized security mechanisms, directly embedding cryptocurrency oriented blockchain schemes into resource-constrained Internet of Video Things (IoVT) networks at the edge is not feasible. By leveraging Electrical Network Frequency (ENF) signals extracted from multimedia recordings as region-of-recording proofs, this paper proposes EconLedger, an ENF-based consensus mechanism that enables secure and lightweight distributed ledgers for small-scale IoVT edge networks. The proposed consensus mechanism relies on a novel Proof-of-ENF (PoENF) algorithm where a validator is qualified to generate a new block if and only if a proper ENF-containing multimedia signal proof is produced within the current round. The decentralized database (DDB) is adopted in order to guarantee efficiency and resilience of raw ENF proofs on the off-chain storage. A proof-of-concept prototype is developed and tested in a physical IoVT network environment. The experimental results validated the feasibility of the proposed EconLedger to provide a trust-free and partially decentralized security infrastructure for IoVT edge networks.

Keywords: electrical network frequency (ENF); Proof-of-ENF (PoENF); consensus; blockchain; security; Internet of Video Things (IoVT)

Citation: Xu, R.; Nagothu, D.; Chen, Y. EconLedger: A Proof-of-ENF Consensus Based Lightweight Distributed Ledger for IoVT Networks. *Future Internet* 2021, *13*, 248. https://doi.org/10.3390/fi13100248

Academic Editors: Ahad ZareRavasan, Taha Mansouri, Michal Krčál and Saeed Rouhani

Received: 7 September 2021
Accepted: 22 September 2021
Published: 24 September 2021

Publisher's Note: MDPI stays neutral with regard to jurisdictional claims in published maps and institutional affiliations.

Copyright: © 2021 by the authors. Licensee MDPI, Basel, Switzerland. This article is an open access article distributed under the terms and conditions of the Creative Commons Attribution (CC BY) license (https://creativecommons.org/licenses/by/4.0/).

1. Introduction

Thanks to the rapid advancements in artificial intelligence (AI) and Internet of Things (IoT) technologies, the concept of Smart Cites becomes realistic. The information fusion capability provided by these interconnected devices enables situational awareness (SAW), which is essential to ensure a safe and sustainable urban environment. With wide deployment of the exponentially increasing smart Internet of Video Things (IoVT) for safety surveillance purposes, intelligent online video stream processing is becoming one of the most actively researched topics in smart cites [1].

In typical Internet of Video Things (IoVT) systems, a huge amount of raw video data collected by geographically scattered cameras is sent to a remote cloud for aggregation. It provides a broad spectrum of promising applications, including public space monitoring, human behavior recognition [2], and suspicious event identification [3]. However, centralized IoVT solutions suffer from the risk of single points of failure and are not scalable for accommodating the ever growing IoVT networks, which are pervasively deployed with heterogeneous and resource-limited smart devices at the edge of networks. Moreover, online video streams and other offline data, such as situation contextual features, are shared among participants using high-end cloud servers, which are under the control of third-party entities. Such a centralized architecture also raises severe privacy and security concerns that data in storage can be misused or tampered with by dishonest entities.

Evolving from the distributed ledger technology (DLT), blockchain has gained significant attention for its potential to revolutionize multiple areas of the economy and society. The inherent security guarantees of blockchain lay down the foundations of serverless record keeping, without the need for centralizing trusted third-party authorities [4]. Blockchain runs on a decentralized peer-to-peer (P2P) network in order to securely store and verify data without relying on a centralized trust authority. The decentralization removes the risk of singular point of failures and mitigates bottleneck performances, which were inherent in centralized architectures. In addition, blockchain leverages distributed consensus protocols to enable a verifiable process for fault tolerance and tamper-proof storage on a public distributed ledger. Therefore, transparency, immutability, and auditability guaranteed by blockchain ensure resilience, correctness, and provenance for all data sharing among untrusted participants.

Internet of Video Things (IoVT) provides a broad spectrum of applications, particularly in the area of public safety [5]. Migrating from centralized cloud-based paradigms to decentralized blockchain-based methods renders IoVT systems more efficient, scalable, and secure. However, directly integrating cryptocurrency-oriented blockchains into resource constrained IoVT systems is difficult in terms of handling the blockchain trilemma [6], which points out that decentralization, scalability, and security cannot perfectly co-exist. Most IoVT devices are highly resource constrained. Therefore, computing and storage intensive consensus protocols are not affordable, such as Proof-of-Work (PoW) [7], Proofs-of-Retrievability (PoRs) [8], or Practical Byzantine Fault Tolerant (PBFT) [9], which come with high communication complexity and poor scalability. In addition, IoVT systems involve a large volume of real-time transactions. Higher throughput and lower latency become key metrics in blockchain-based systems for IoVT deployed on edge networks. Furthermore, DLTs are not general-purpose databases. The storage overhead is prohibitively high if raw data generated by IoVT transacting networks are stored in the blockchain.

The Electrical Network Frequency (ENF) is the power supply frequency which fluctuates around its nominal frequency (50/60 Hz). The frequency fluctuations vary based on geographical region. The ENF fluctuations estimated from simultaneously recorded audio/video recordings within a power grid have a high correlation similarity [10].

Inspired by spatio-temporal sensitive ENF contained in multimedia signals, this paper proposes *EconLedger*, a novel *Proof-of-ENF* (PoENF) consensus algorithm based lightweight DLT for small scale IoVT networks. Compared to PoW or PoRs, which require high computation or storage resources in mining process, our novel PoENF consensus requires each validator to use extracted ENF variations from simultaneous multimedia recordings as proofs during current consensus round. The validator that presents a valid ENF proof with minimal squared-distance-based score is qualified to generate a new block. Thus, the PoENF consensus mechanism not only achieves efficiency without high demand of mining resource or hardware platform support but it also enhances security by mitigating mining centralization.

In contrast to existing solutions that directly collect ENF fluctuations from power grids and stores audio/video recordings in a centralized location-dependent ENF database [10,11], EconLedger uses *Swarm* [12], which is a decentralized database (DDB) technology, to archive raw ENF-containing multimedia proofs and transactions over IoVT networks. Only hashed references of data are recorded on an immutable and auditable distributed ledger. Thus, it reduces the ever-increasing data storage overhead on the public ledger. The EconLedger ensures correctness, availability, and provenance of data sharing among untrusted devices under a distributed network environment. Moreover, a network with permission ensures that only authorized nodes can access raw data on DDB such that privacy preservation is guaranteed.

In summary, this paper makes the following contributions:

(1) A secure-by-design EconLedger architecture is introduced along with detailed explanation of the key components and work flows;
(2) A novel PoENF consensus mechanism is proposed, which improves resource efficiency and achieves a higher throughput than PoW-based blockchains;

(3) A finalized on-chain ledger is coupled with a decentralized off-chain storage to resolve storage burden, and it guarantees security and robustness of data sharing and cooperation in IoVT networks;

(4) A proof-of-concept prototype is implemented and tested on a small scale IoVT network, and experimental results verified that the EconLedger is feasible and affordable with respect to the IoVT devices deployed at edge networks.

The remainder of this paper is organized as follows: Section 2 briefly discusses background knowledge of ENF, then reviews existing consensus algorithms and state-of-the-art research on IoT Blockchains. Section 3 introduces the rationale and architecture of EconLedger, as well as core features and security guarantees. A novel PoENF consensus mechanism is explained in Section 4. Section 5 presents prototype implementation and numerical results and discusses performance improvements and security insurances. Finally, a summary is presented in Section 6.

2. Background and Related Work

This section introduces how ENF can be generated from multimedia streams and how ENF can be used for the environmental fingerprint. Following that, we describe typical consensus protocols in blockchain and provide related work on IoT-blockchain integration.

2.1. ENF as a Region-of-Recording Fingerprint

ENF is the supply frequency in power distribution grids, which has a nominal frequency of 50 Hz or 60 Hz depending on the location of the power grids. Due to environmental effects in the grid such as load variations and control mechanisms, the instantaneous ENF usually fluctuates around its nominal value. At a given time, variation trends of ENF fluctuations from all locations of the same grid are almost identical due to the interconnected nature of the grid [13]. ENF fluctuations are embedded in audio/video recordings either due to electromagnetic induction or background hum from devices connected to the power grid [14]. Thanks to the consistency and reliability of ENF at a time instant, ENF has been adopted as a forensic tool for identifying forgeries in multimedia recordings. All ENF signals estimated from simultaneous multimedia recordings at different locations have similar fluctuations throughout the power grid. Thus, there are multiple forensic applications based on ENF, such as validating the time-of-recording of an ENF-containing multimedia signal [14] and estimating its location-of-recording [15].

In IoVT systems, ENF signals extracted from video recordings are in the form of illumination frequency (120 Hz). The video recordings made under indoor artificial light include ENF fluctuations. The estimation of ENF signals depends on the type of imaging sensor used in a camera. The most commonly used imaging sensors are complementary metal oxide semiconductors (CMOSs) and charge-coupled device (CCD) sensors, which have different shutter mechanisms. In this work, we assume that ENF signals are extracted from video recordings generated by cameras with CMOS imaging sensors in an indoor setting with artificial light [11,16]. The estimation of ENF involves various signal processing techniques such as power spectral analysis and spectrogram-based techniques, which are beyond the scope of this paper.

2.2. Consensus Protocols for Blockchain

This section introduces consensus protocols regarding diverse blockchain networks that are typically classified into permissionless blockchain (e.g., Nakamoto protocol) or permissioned blockchain (e.g., PBFT).

2.2.1. Nakamoto Protocols

The Nakamoto protocol is implemented as the consensus foundation of Bitcoin [7], and it is widely adopted by many cryptocurrency-based blockchain networks such as Ethereum [17]. The Nakamoto protocol adopts a computation-intensive PoW, which requires all participants to compete for rewards through a cryptographic block-hash value discovery racing game. The PoW consensus demonstrates security and scalability in an asynchronous open-access network as long as an adversary does not control the majority (51%) of the miners. However, the brute-force PoW mining process also incurs a high demand in terms of computation and energy consumption such that it is not affordable on resource-constrained IoT devices.

In order to improve performance and resource usage efficiency in PoW, a number of alternative Proof of X-concept (PoX) schemes have been proposed. Permacoin [8] repurposes mining resources in PoW to achieve distributed storage of archival data. The Permacoin adopts PORs [18], which require miners to present random access to a copy of a file from local storage as valid proof for successfully minting money. Permacoin requires participants to invest in its storage capacity rather than solo computational power. It could reduce unnecessary wastage of computational resources in PoW and mitigate centralized mining pools issue.

Similar to Permacoin, a Resource-Efficient Mining (REM) [19] scheme is proposed to achieve security and resource efficiency based on the partially decentralized trust models inherent in Intel Software Guard Extensions (SGX). The REM utilizes a Proof-of-Useful-Work (PoUW) consensus protocol, which requires miners to provide trustworthy measurements on CPU cycles used by its useful workloads in SGX-protected enclave. Compared with Proof-of-Elapsed-Time (PoET) in Sawtooth [20] that uses random idle CPU time as proofs, PoUW in REM not only prevents the stale chip problem but also yields the smallest amount of mining waste.

In order to reduce energy consumption caused by intensive hash value calculating in PoW, Peercoin [21] adopts Proof-of-Stake (PoS), which leverages the distribution of token ownership to simulate a verifiable random function to propose new blocks. Such a process of efficient "virtual mining" manner allows PoS miners to only consume limited computational resources in order to generate new blocks. Similarly to PoW, PoS guarantees security as long as an adversary owns no more than half of the total stakes in the network.

Unlike PoW and its variants, the PoENF consensus scheme neither requires high demand of computation and storage for mining nor depends on security guarantees supported by trusted hardware or monetary deposit stake. It is suitable for heterogeneous IoVT devices connected to the power grid.

2.2.2. Byzantine Fault Tolerant Protocols

As the first practical BFT consensus, PBFT [9] uses the State Machine Replication (SMR) scheme to address the Byzantine General Problem [22] in distributed networks. It has been widely adopted as a basic consensus solution in the permissioned blockchains, such as Hyperledger Fabric [23]. The PBFT algorithm guarantees both liveness and safety in synchronous network environments if at most $\lfloor \frac{n-1}{3} \rfloor$ out of total of n replicas are Byzantine faults. Compared to the probabilistic Nakamoto blockchains, BFT-based consensus networks ensure a deterministic finality on distributed ledger. However, it inevitably incurs high latency and communication overhead as synchronously executing consensus protocol among all nodes in large scale networks.

Therefore, combining Nakamoto-style block generation with BFT-style chain finality provides a prospective solution to ensure data consistency and immediate finality. Casper [24] introduces a lightweight chain finality layer on top of a Nakamoto protocol, similarly to PoW and PoS. In Casper, a fixed set of validators executes a PoW block proposal protocol to maintain an ever-growing *block tree*, while an efficient voting-based process is responsible to commit a direct ancestor block of the finalized parent block as a *checkpoint*. Finally, only a unique checkpoint block path from checkpoint tree is accepted as the finalized chain.

Unlike Casper, which is a PoS-based finality system overlaying an existing PoW blockchain, our EconLedger uses a voting-based chain finality in order to resolve the forks caused by probabilistic PoENF block generation.

2.3. State of the Art on IoT-Blockchain

To support security and lightweight features required in IoT systems, the IoTChain [25] proposes a three-tier blockchain-based IoT architecture, which allows regional nodes to perform any lightweight consensus, such as PoS and PBFT. IoTChain only provides simulation results on communication cost of transactions; however, key metrics in the consensus layer, such as computation, storage, and throughput, are not considered. FogBus [26] proposes a lightweight framework for integrating blockchain into fog-cloud infrastructure, which aims to ensure data integrity as transferring confidential data over IoT-based systems. In FogBus, master nodes deployed at the fog layer are allowed to perform PoW mining, while IoT devices send transactions to master nodes as trust intermediates to interact with blockchain. However, using PoW as the backbone consensus protocol still results in high energy consumption and low throughout.

HybridIoT [27] proposes hybrid blockchain-IoT architecture in order to improve scalability and interoperability among sub-blockchains. In HybridIoT, a BFT inter-connector framework functions as a global consortium blockchain to link multiple PoW sub-blockchains. However, using PoW consensus in sub-blockchain networks still imports computation and storage overhead on IoT devices if they are deployed as full nodes. IoTA [28] aims to enable cryptocurrency designed for the IoT industry, and it leverages a directed acyclic graph (DAG), called tangle [29], to record transactions rather than chained structure of the ledger. IoTA provides a secure data communication protocol and zero fee micro-transaction for IoT/machine-to-machine (M2M), and it demonstrates high throughput and good scalability. However, existing IoTA networks still rely on hard-coded coordinators, which employ PoW to finalize the path of recorded transactions in DAG.

Unlike the above mentioned IoT-Blockchain solutions, which either adopt computation intensive PoW as their backbone consensus mechanism or rely on an intermediate fog layer to execute consensus protocol, EconLedger aims to provide a partially decentralized and lightweight blockchain for resource constrained IoVT devices at the edge without relying on any intermediate consensus layer deployed at fog level. Moreover, EconLedger leverages DDB technology to enable trusted off-chain storage, which reduces storage overhead caused by directly storing raw data on the public distributed ledger.

3. EconLedger: Rationale and Architecture

This section provides a comprehensive overview of EconLedger system architecture consisting of the following: (1) upper-level IoVT application layer; and (2) Econledger fabric enabled security networking infrastructure. Following that, we explain the network model of EconLedger with basic security assumptions and describe an efficient hybrid on-chain and off-chain storage structure based on the DDB system.

3.1. System Design Overview

EconLedger aims at a secure-by-design, trust-free and partially decentralized infrastructure for cross-devices networking IoVT systems at the edge. We consider a small scale video surveillance network with 100 nodes, and all IoVT devices and edge/fog servers are connected to the same regional power grid. Here, a node refers to a device owned by a user. Figure 1 is the system architecture of EconLedger.

3.1.1. IoVT Application

The upper-level IoVT application utilizes an EconLedger fabric to enable decentralized video analytic services and information visualization at the edge. All devices and users must be registered to join the IoVT system as required by the permissioned network, which can provide basic security primitives such as public key infrastructure (PKI), identity authentication [30], and access control [31], etc. Real-time video streams generated by cameras are transferred to on-site/near-site edge devices for lower level analytic tasks, such as object detection and situational contextual features extraction. Thus, cameras associated with edge devices act as IoVT service units at the network of edge. Then, IoVT service units send raw video data and extracted contextual information to the information visualization unit, which provides video recordings and smart applications for authorized users.

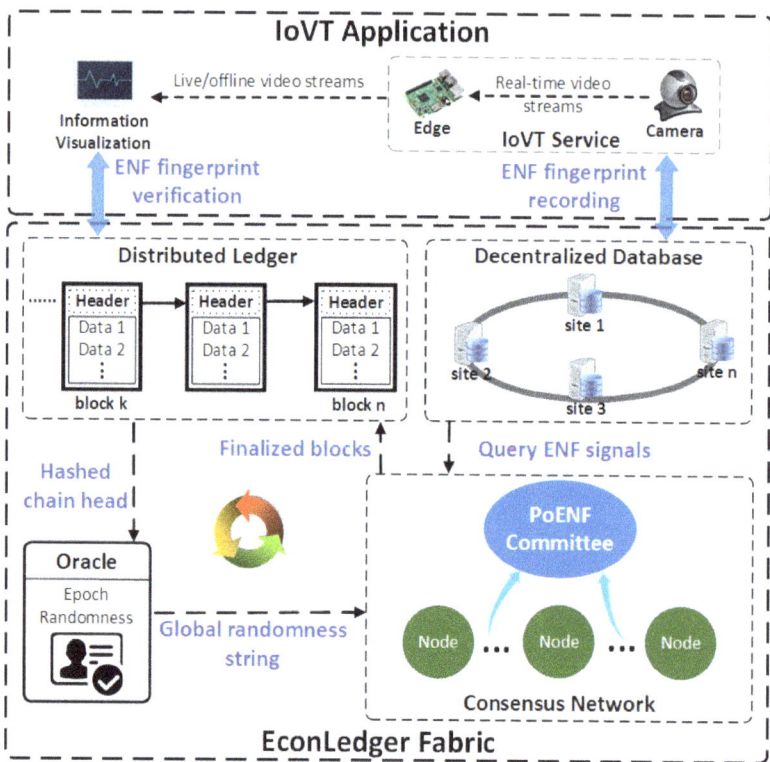

Figure 1. The EconLedger system architecture.

To prevent visual layer attacks, IoVT service extracts ENF signals from video streams as an environmental fingerprint, which is stored into DDB and secured by *EconLedger fabric*. At any given time instant, variation trends of ENF-containing multimedia signals from all synchronous cameras on the same power grid are almost identical. Therefore, using ENF fluctuations recorded on EconLedger laid solid ground truth for video authenticity verification. By calculating correlation coefficients among ENF signals extracted from video recordings with an agreed ENF estimate recorded on distributed ledger, the information visualization unit verifies whether or not live/offline video streams are generated by cameras within the same power grid [32,33].

3.1.2. EconLedger Fabric

The EconLedger fabric provides fundamental networking and security infrastructure to support decentralized security features for the IoVT system. All authorized devices firstly store raw ENF fingerprints into the DDB, then the devices launch transactions that include hashed references of raw data along with valid signatures. As transactions store fixed-length hashed references rather than raw data with varying size, such an off-chain manner reduces storage overhead when IoVT devices verify transactions and synchronize the ever-increasing distributed ledger.

EconLedger uses a small PoENF committee to achieve high efficiency by reducing message propagation delay and communication overhead on the edge network. Given a random committee election mechanism, only a subset of nodes within the network are elected as PoENF committee members. The PoENF consensus protocol is only executed by validators of a PoENF committee instead of all nodes in the network. Therefore, scalability is improved at the cost of partial decentralization by a PoENF consensus committee.

Meanwhile, a random PoENF committee rotation strategy ensures that robustness is not sacrificed due to fewer validators. Combining the current status of the distributed ledger, a distributed randomness protocol acts as the oracle to periodically generate global randomness strings for PoENF committee selection. As randomness strings are bias-resistant and unpredictable, the probability of an adversary dominating a subsequent committee decreases exponentially even if the current PoENF committee is compromised.

3.2. Network Model

EconLedger relies on a permissioned network, and we assume that the system administrator is a trust oracle for maintaining global identity profiles for all valid nodes. We adopt a standard asymmetrical algorithm such as Rivest–Shamir–Adleman (RSA) for key generation (RSA.gen) and digital signature scheme (RSA.sign, RSA.verify). During the registration process, signing-verification key pair $(sk_i, pk_i) \leftarrow RSA.gen(i)$ is generated by PKI and assigned to the authorized node u_i. Additionally, a node's public key pk_i is associated with its credit stake $c_i \leq C_{max}$, where C_{max} is the maximum value of credit stake defined by the system. Therefore, all registered nodes can be represented as $U = \{(pk_1, c_1), (pk_2, c_2), ..., (pk_n, c_n)\}$, where n is the total number. As the above security assumptions depend on the system administrator's behavior, our EconLedger is a partially decentralized blockchain model.

EconLedger assumes a synchronous network environment. Operations in consensus protocol are coordinated in rounds with upper bounded delay T_Δ. Thus, the time is divided into discrete *slots*, which can be indexed by logical clocks *ticks* to synchronize the events in a distributed system [34]. Given a certain tick $t \in \{1, 2, 3, ...\}$, slot sl_t represents the length of time window to measure T_Δ. The time window of sl_t should be sufficient to guarantee that the message transmitted by a sender is received by its intended recipients (accounting for local time discrepancies and network delays). Thus, we require $sl_t \geq T_\Delta$ in order to ensure the liveness of consensus protocol.

3.3. Hybrid On-Chain and Off-Chain Storage

To address issues of high storage overhead incurred by directly saving raw data into DLTs, EconLedger utilizes a hybrid on-chain and off-chain storage solution. Figure 2 illustrates the block and off-chain data structure used in EconLedger. The block is the basic unit of on-chain storage, which includes block header and the orderly transactions list. The MT_root in the block header stores the hash root of a Merkle tree to maintain the integrity of all transactions. In each transaction, the *swarm_hash* only stores references to the data rather than the data themselves. As references are hash values with fixed length such as 32 or 64 bytes, all transactions have almost the same size even if linked raw data have large sizes or require different formats, such as ENF signals or multimedia recordings.

Off-chain storage relies on a Swarm network in which all sites cooperatively construct a DDB system. In EconLedger, a site refers to a fog/edge server. The data uploaded to Swarm are cut into pieces called *chunks*, which is the basic unit of storage and retrieval in the Swarm network. Each chunk can be accessed at a unique address, which is calculated by its hashed content. All data chunks use their chunk hash to construct a Merkel hash tree for which its root is the reference to retrieve raw data. Swarm implements a specific type of content addressed distributed hash tables (DHTs), called Distributed Pre-image Archive (DPA), to manage chunks across distributed sites. All Swarm sites have their own base addresses with the same size as the chunk hash, and the sites closest to the address of a chunk not only serve information about the content but actually host data [35]. All sites in the Swarm network use the Kademlia DHT protocol [36], which synchronizes chunks in a P2P manner, to ensure data persistence and redundancy.

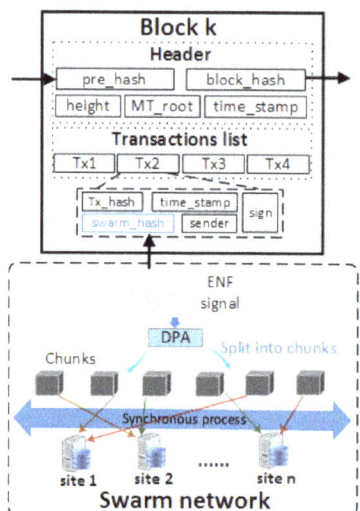

Figure 2. The illustration of block and off-chain data structure.

4. PoENF: A Proof-of-ENF Consensus Protocol

In this section, basic notations used in protocol design are clearly defined and explained. Then, an overview of PoENF consensus protocol is illustrated so that the reader can understand key components and workflow. Following that, we offer details on Byzantine resistant PoENF algorithms in block generation along with a voting-based chain finality. Finally, we also describe incentive mechanisms including rewards and punishment strategies given by mathematical analysis.

4.1. Basic Notation

Table 1 describes relevant notation used in PoENF model. To model sequential events in synchronous consensus rounds, a set of subsequential slots are used to define *Epoch*, which is represented as $sl_E = \{sl_1, sl_2, ..., sl_t\}$, where $0 \leq t \leq R$, and epoch size R is a value of multiple unit slot sl. A *validator* $v_i \in V$ ($V \subseteq U$) is a valid node that is qualified for being selected as a PoENF committee member. We define *Dynasty* to represent current PoENF committee, which is denoted as $D = \{(pk_1, c_1), (pk_2, c_2), ..., (pk_k, c_k)\} \subseteq V\}$, where $0 \leq k \leq K$, and K is the PoENF committee size. We use $\mathcal{H}(\cdot)$ to denote a predefined collision-resistant hash function that outputs hash string $h \in \{0,1\}^\lambda$.

Table 1. Relevant basic notation.

Symbol	Description
sl_E	Epoch including sequential order of time slot
D	Dynasty represents current PoE committee
tx	A transaction broadcasted by the node of network
B	A block proposed by the validator in current Dynasty
C	Distributed ledger maintained by the consensus network

Before introducing key features and components in the PoENF consensus protocol, several basic definitions are introduced as following.

Definition 1. Transaction is used to save data that are launched by a node u_i for recording on the distributed ledger, and its structure is represented as $tx = \{tx\_hash, pk_i, T_{stamp}, data, \sigma_i\}$, where the parameters are the following:

- tx_hash is a λ-bit-length hash string of transaction tx, which is calculated by $\mathcal{H}(pk_i, T_{stamp}, data)$;
- pk_i is sender's public key;
- T_{stamp} is time stamp of generating transaction;
- $data$ is the information $d \in \{0,1\}^*$ enclosed by the transaction, such as swarm_hash or any byte strings;
- σ_i is a signature $RSA.sign_{sk_i}(tx\_hash, pk_i, T_{stamp}, data)$ signed by the sender's private key sk_i.

Definition 2. Block is a basic data unit that encapsulates valid transactions and is always appended on the chain head. A block generated at slot sl_t ($t \in 1,2,3,..$) by validator v_j is represented by $B_i = (pre\_hash, height, mt\_root, tx\_list, sl_t, pk_j, \sigma_j)$, where the parameters are the following:

- pre_hash is a λ-bit-length hash string of previous Block B_{i-1}, which is calculated by $\mathcal{H}(pre\_hash, height, mt\_root, tx\_list, sl_{t-1})$;
- $height$ is the height of current block in blockchain (ledger);
- mt_root is a root hash of a Merkle tree of tx_list;
- tx_list is an orderly transactions list $[tx_1, tx_2, ..., tx_n]$;
- sl_t is a block created time stamp at the round sl_t;
- pk_j is public key of validator v_j;
- σ_j is a signature $RSA.sign_{sk_j}(pre\_hash, height, mt\_root, tx\_list, sl_t, pk_j)$ signed by validator v_j.

We define a special block called *Genesis Block* that is represented as $B_0 = (pre\_hash = 0, height = 0, sl_0 = 0, init\_D)$, where $init\_D$ is the initial dynasty. Therefore, all on-chain data on the distributed ledger are organized as an ordered sequence of blocks starting from B_0.

Definition 3. Blockchain (Distributed Ledger) is a partial order of blocks that is represented as $\mathcal{C} = B_0 \to B_1 \to ... \to B_{n-1} \to B_n$ indexed by strictly increasing slots sl_t. Each block B_i uses its $pre\_hash=\mathcal{H}(B_{i-1})$ to link with the previous block B_{i-1}, and key parameters are the following:

- length: the length of the chain denoted $len(\mathcal{C}) = n$ to count the number of blocks between the genesis block B_0 and the confirmed block B_n;
- head: the head of the chain denoted $head(\mathcal{C}) = B_n$, where B_n is the last confirmed block that is extended on finalized main chain.

4.2. PoENF Committee Consensus Protocol: Overview

Figure 3 is an overview of the PoENF consensus protocol, which includes the distributed ledger structure and PoENF committee consensus workflows. The distributed ledger in EconLedger follows a tree structure originated from the genesis block. Each new block extends its chain path through *pre_hash* to point to a parent block. All nodes in such a ledger tree can be represented as confirmed blocks (blue) or finalized blocks (red), as the upper part of Figure 3 shows. The chain height follows a strictly increasing sequence of finalized blocks; therefore, a valid path can only proceed through those red nodes. The head of a blockchain is anchored on a recently confirmed block that has linked to a finalized chain with the largest height value.

At the configuration stage, the system administrator specifies a group of validators as the initial PoENF committee to initialize an EconLedger network. The lower part of Figure 3 demonstrates workflows of the PoENF committee consensus protocol, including the dynasty cycle and the epoch cycle. A dynasty cycle starts from committee selection and ends when the global randomness string of current dynasty has been updated by the randomness change process. At the beginning of a dynasty's lifetime, the committee selection process uses the current global randomness string as a seed for committee election protocol, which exploits a Verifiable Random Function (VRF) based cryptographic sorting scheme [37]. Given the credit weights of all nodes, K validators are randomly chosen to construct a new PoENF committee D, which will be added to the current block. Finally, validators of new D establish a fully connected P2P consensus network and start a new dynasty cycle.

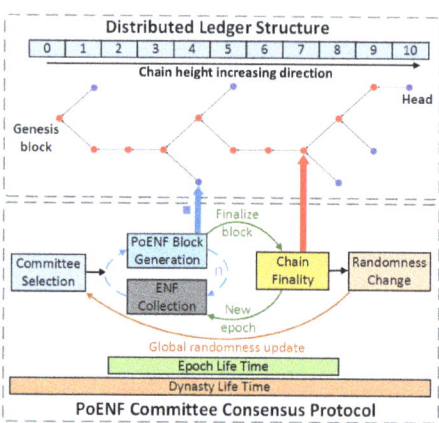

Figure 3. The PoENF consensus protocol overview.

For each epoch cycle, PoENF block generation and chain finality are core functions that ensure liveness and termination in continuous consensus rounds. By calculating a squared distance of ENF proofs from validators, the PoENF mechanism determines that a validator with the minimal squared-distance-based score can generate a valid block in the current block proposal round. After n rounds of block proposal, chain finality relies on a voting-based chain finality mechanism to resolve fork issues caused by conflicting confirmed blocks and finalizes the history of ledger data by using a unique chain path.

At the end of current dynasty, PoENF committee members utilize the RandShare mechanism to cooperatively reach agreement on proposing a new global randomness string. As a distributed randomness protocol, RandShare adopts Publicly Verifiable Secret Sharing (PVSS) [38] to ensure unbiasability, unpredictability, and availability in public randomness sharing. The proposed unbiasable and unpredictable global randomness string will be updated as the new seed for the committee selection process of the subsequential dynasty.

4.3. PoENF-Based Block Proposal Mechanism

The PoENF-based block proposal mechanism is mainly responsible for generating candidate blocks and extending them along a finalized chain path. Following the principles of chain-based Nakamoto protocols, the PoENF algorithm simulates a virtual mining method by pseudorandomly specifying a validator of committee as the slot leader to generate a block. To generate a block, a validator must present an ENF proof that has the minimum squared distance score in current round. All honest validators accept valid blocks and ensure that only one block is extended on the finalized main chain of local distributed ledger.

4.3.1. Transactions Pooling

Given a certain period of sliding window for ENF collection, each validator collects transactions from valid nodes. If a transaction stores reference that point to ENF proof, it is an ENF proof transaction. In the current block generation round, each node is required to send only one ENF proof transaction. After receiving the broadcasted transactions, each validator verifies buffered transactions according to predefined conditions:

(i) Transaction sender $u_i \in U$ and $RSA.verify(tx\_hash, pk_i, T_{stamp}, data) = \sigma_i$ by using the sender's public key pk_i;
(ii) ENF proof tx sent by u_i should not exist in the transactions pool;
(iii) Time stamp T_{stamp} must fall into current time slot.

The condition (i) prevents transactions from invalid nodes or any malicious modification, while conditions (ii) and (iii) are mainly for preventing duplicated ENF proof tx in current period time slot. After the verification process, only valid transactions are cached as local transaction pools denoted as $TX = \{tx_1, tx_2, ..., tx_N\}$, where N is the transactions pool size. The validator also uses condition (iii) to regularly check the local transaction pool and removes outdated transactions that have not been recorded in the latest confirmed block.

4.3.2. PoENF Consensus Algorithm

Given current transactions pool TX, the validator $v_i \in D$ chooses all ENF-proof transactions generated by committee members to construct an ENF-proof transactions list $TX_{ENF} = \{tx_1, tx_2, ..., tx_K\}$, where K is the committee size. An ENF proof is a vector $E = \{e_1, e_2, ..., e_d\}$, where $e_i \in \mathbb{R}$ is the ENF sample value, and d is the samples' size. By using $swarm\_hash$ that is stored in the $data$ parameter of tx_k, the E_k sent by v_k can be fetched from off-chain storage. Thus, each v_i can locally maintain a set of collected ENF proof vectors $G_i = \{E_1, E_2, ..., E_k\}$, where $k \leq K$.

In order to become a slot leader and propose a new block in the current block proposal round, v_i must show that its ENF proof E_i can solve a PoENF puzzle problem. Intuitively, the goal of PoENF puzzle problem is to choose the E_k that deviates the least from all ENF proofs in G_i based on their relative distances, which are computed with the Euclidean norm. However, a single Byzantine validator can force the PoENF algorithm to choose any arbitrary ENF proof by sending a poisoned E_b that is too far away from other ENF proofs. Therefore, our PoENF algorithm adopts the Krum aggregation rule to provide the (α, f)-Byzantine resilience property [39].

For each $v_i \in D$, let $G_i = \{E_1, E_2, ..., E_n\}$ include $n \geq 2f + 3$ collected ENF proofs from PoENF committee members, and at most only f is sent by Byzantine nodes. For any $i \neq j$, let $i \to j$ denote the fact that E_j belongs to the $n - f - 2$ closest ENF proofs to E_i. Then, we define the ENF score for v_i.

$$s(i) = \sum_{i \to j} \|E_i - E_j\|^2. \quad (1)$$

Equation (1) calculates ENF scores $(s(1), ..., s(n))$ associated with validators v_1 to v_n, respectively, and applies the Krum rule to select the minimum ENF score as follows.

$$s^* = \underset{i \in \{1,...,n\}}{\operatorname{argmin}} (s(i)). \quad (2)$$

Finally, the PoENF puzzle problem is formally defined as the following.

Definition 4. *Proof-of-ENF: Given $G_i = \{E_1, E_2, ..., E_n\}$ collected by validator $v_i \in D$, the process of PoENF verifies whether a valid ENF proof E_j can meet the condition $s(j) \leq s^*$. If it does, v_j wins the leader election and is qualified to propose a block; otherwise, the blocks generated by any v_j are rejected.*

Given the above definitions, the PoENF-enabled block generation procedures are presented in Algorithm 1. During the current block generation round slot sl_t, each validator $v_i \in D$ executes the *generate_block()* function to propose a candidate block according to its collected ENF proofs in the transactions pool. If the ENF score s_i is not greater than the target value s^*, then v_i can create a new block and broadcast it to the network. Otherwise, they are only allowed to verify blocks from other validators until the current round finished. The closer s_i is to s^*, the higher probability that v_i can propose a new block.

Algorithm 1 The PoENF-based block generation procedures.

1: **procedure:** generate_block(v_i)
2:    $hc \leftarrow \mathcal{H}(head(\mathcal{C}))$
3:    $height \leftarrow head(\mathcal{C}).height + 1$
4:    $mt\_root \leftarrow \text{MTree}(v_i.TX)$
5:    $[E_1, E_2, ..., E_n] \leftarrow \text{ENF\_vect}(v_i.TX)$
6:    $enf\_score \leftarrow [\,]$
7:    **for** E_i in $[E_1, E_2, ..., E_n]$ **do**
8:       $s_i \leftarrow \sum_{i \to j} \|E_i - E_j\|^2$
9:       $enf\_score.append(s_i)$
10:   **end for**
11:   $s^* \leftarrow \text{Min}(enf\_score)$
12:   **if** $s_i \leq s^*$ **then**
13:      $new\_block \leftarrow (hc\|mt\_root\|v_i.TX\|v_i.pk\|sl.t\|height)$
14:      $\sigma_i \leftarrow \text{Sign}(new\_block, v_i.sk)$
15:      **return** $(new\_block\|\sigma_i)$
16:   **end if**
17: **procedure:** verify_block(new_block, σ_j)
18:   **if** Verify_Sign(new_block, σ_j) $\neq$ True OR
19:      Verify_TX(new_block) $\neq$ True **then**
20:      **return** False
21:   **end if**
22:   $hc \leftarrow \mathcal{H}(head(\mathcal{C}))$
23:   **if** $new\_block.height \neq head(\mathcal{C}).height + 1$ OR
24:      $new\_block.hc \neq hc$ **then**
25:      **return** False
26:   **end if**
27:   $[E_1, E_2, ..., E_n] \leftarrow \text{ENF\_vect}(new\_block.tx\_list)$
28:   $enf\_score \leftarrow [\,]$
29:   **for** E_i in $[E_1, E_2, ..., E_n]$ **do**
30:      $s_i \leftarrow \sum_{i \to j} \|E_i - E_j\|^2$
31:      $enf\_score.append(s_i)$
32:   **end for**
33:   $s^* \leftarrow \text{Min}(enf\_score)$
34:   **if** $s_j > s^*$ **then**
35:      **return** False
36:   **end if**
37:   **return** True

In block verification process, each validator calls the $verify\_block()$ function to determine whether or not the received new_block can be accepted and appended to chain. The $Verify\_Sign()$ checks if a new_block sent by v_j has a valid signature σ_j, while $Verify\_TX()$ validates that all transactions recorded in new_block are sent from valid nodes and have the same Merkle tree root as $new\_block.mt\_root$. After validating that new_block is generated in the current round slot sl_t with the correct chain header, a PoENF algorithm verifies if v_j has a valid ENF proof with minimum score. If all conditions are satisfied, the new_block is accepted into confirmed status, and v_i updates the head of local chain as $head(\mathcal{C}) = B_{i+1}$ accordingly. Otherwise, the new_block is rejected and discarded.

4.3.3. Chain Extension Policies

In a PoENF block generation round, validators extend the local chain based on a *"largest height of confirmed block"* rule, which requires that new blocks B_{i+1} are only appended to $head(\mathcal{C})$ by letting $B_{i+1}[pre\_hash] = \mathcal{H}(head(\mathcal{C}))$. The PoENF process allows that the probability of a block generated by validator v_i is related to the rank of its ENF proof E_i among the global ENF proof vectors G. However, G_i observed by validator v_i may vary due to network latency or misbehavior of Byzantine nodes. Thus, it is hard to guarantee that only one block is proposed during each slot round, and the amount of candidate blocks could be between zero and the committee size K. Given the candidate blocks number $b \in [0, K]$, the chain extension rules are described as follows:

(i) $b = 1$: If there is only one proposed candidate block B_{i+1}, then the block is accepted as confirmed status and updates the chain head as $head(\mathcal{C}) = B_{i+1}$.

(ii) $b > 1$: If more than one candidate blocks are proposed, then all blocks are accepted as confirmed status. The $head(\mathcal{C})$ update follows two sub-rules:

  (a) Chain head points to a block that records most ENF proof transactions among other blocks;
  (b) For the candidate blocks having ENF proof transactions of the same size, the block generated by validator that has the largest credit value becomes the chain head.

(iii) $b = 0$: If none of the validators proposes a block at the end of current slot round, block generation follows a spin manner. As validators of current committee can be sorted by account address, we can calculate $ind = height \pmod{K}$. Thus, a validator at rank ind can also propose a candidate block in the current round. The chain head update process follows the rule i) $b = 1$.

Rule (i) covers a basic scenario to ensure that a new blocks is extended on the chain head. Rule (ii) handles the conflicting chain head update scenario when multiple validators propose valid blocks when they have different global ENF proof vectors G. It also discourages dishonest behaviors by using a smaller G to win the right to block proposal. Rule (iii) guarantees liveness in PoENF consensus process such that at least one uniform block is generated to ensure chain extension even if a leader cannot propose a new block due to crash failures or attacks.

4.4. Voting-Based Chain Finality Mechanism

Since fork issues are caused by network latency or deliberate attacks, the block proposal mechanism will inevitably produce multiple conflicting blocks, which are children blocks with the same parent block. Therefore, those proposed blocks are in fact form an ever-growing *block tree* structure, as the upper part of Figure 3 shows. At the end of an epoch, the *head* with epoch height becomes a checkpoint that is used to resolve forks and to finalize chain history. Inspired by Casper [24] and Microchain [40], our EconLedger finality process adopts a voting-based finality mechanism overlaying the PoENF block generation to commit checkpoint block and finalize the already committed blocks on the main chain.

The chain finality protocol is mainly for identifying a unique chain path on block tree by choosing a single child block from multiple children blocks with a common parent block. For efficiency purposes, the chain finality protocol is only executed on *checkpoint* blocks rather than the entire block tree, and committee members vote for hashes of blocks instead of entire block contents. The chain finality ensures that only one path, including finalized blocks, becomes the main chain. Therefore, blocks generated in the new epoch are only extended on such a unique main chain.

4.5. Incentives and Punishment Strategies

Although the contributions of this manuscript include performance improvement and security guarantees by EconLedger, this section briefly discusses incentive design while leaving detailed analysis for future investigation. EconLedger uses an incentive mechanism to reward validators who behave honestly and make contributions in the PoENF block generation and chain finality process. At the end of a block generation cycle, transactions fees included in the confirmed block construct a rewarding fees pool that can be distributed to all validators in the current round. The incentive mechanism uses ENF score to evaluate a validator's contribution, and reward fees that are distributed to $v_i \in D$ are proportional to its ENF score s_i. Let $S = \{s_1, s_2, ..., s_n\}$ denote ENF scores, the reward rule is defined as follows:

$$\gamma_i = \frac{\frac{1}{s_i}}{\sum_i^n \frac{1}{s_i}} \mathcal{R}, \qquad (3)$$

where γ_i is the reward fee that v_i obtains from the total reward fees $\mathcal{R}$ during current block generation round. The smaller the ENF score of a validator, the higher the reward fees it can gain. As the variations of ENF proofs from all honest nodes are trivial during ENF collection time, collected rewarding fees in $\mathcal{R}$ are almost evenly distributed to honest contributors. As ENF fluctuations are randomly generated from power grids and vary at different times, Byzantine nodes can only gain marginal benefits by using duplicated or arbitrary ENF proofs that have large ENF scores.

In addition to rewarding fees, the credit stake c_i of a honest validator v_i will also increase by one as a reputation reward. The higher the credit stake c, the higher the probability that a validator is selected as a PoENF committee member. Unlike PoS, credits in EconLedger are not directly associated with any type of currency, and they are not transferable in any format of transactions. Therefore, all users are encouraged to behave honestly to gain more benefits by increasing their reputation credits. Moreover, credit stake c of a node cannot excel an upper-bounded limitation C_{max}, for instance, no more than 10. Therefore, an adversary cannot simply accumulate its credit stake to achieve mining centralization and then control the majority power of the network.

A punishment strategy is also designed to discourage dishonest behaviors, such as withholding its ENF proof, proposing multiple blocks in current round, or violating chain extension rules. After PoENF committee selection, each $v_i \in D$ must deposit a fixed amount of fees to its *security stake* sc_i. If any misbehaving actions in consensus process v_i are detected, the balance of sc_i will be slashed as punishment. In addition, its credit stake c_i also decreases by one. Given the assumption that an adversary can only compromise no more than f nodes on the network, the slashing security deposit rule can increase financial cost if attackers use these compromised nodes to disturb consensus protocol, while reducing credit stake results in the lower probability that Byzantine nodes can be selected as committee members.

5. Experiment and Evaluation

In this section, a proof-of-concept prototype implementation and experimental configuration is described. Following that, we evaluate Econledger based on numerical results in terms of network latency, computation overhead, and communication throughput. Then, comparative experiments based on benchmark blockchain platforms are performed to show performance improvement. Finally, we analyze the performance and security properties provided by EconLedger.

5.1. Prototype Implementation and Experimental Setup

To verify the proposed EconLedger, a concept-proof prototype is implemented in Python, which consists of approximately 3100 lines of code. We adopted Flask [41], which is a light micro-framework for Python application, in order to implement networking and web service APIs for EconLedger node. All cryptographic functions are developed on the foundation of standard python lib: cryptography [42], such as using RSA for key generation and digital signature and using SHA-256 for all hash operations. As a lightweight and embedded SQL database engine, SQLite[43] is adopted to manage on-chain storage, such as ledger data and peering nodes information.

Table 2 describes the devices used for the experimental study. The prototype is deployed on a small-scale local area network (LAN) that consists of multiple desktops and IoT devices. The prototype of EconLedger emulates an office building setting: a Dell Optiplex-7010 functions as a monitor server to collect data from scattered IoVT services deployed at different locations of the building, while all Raspberry Pi (RPi) boards play the role of edge devices that process raw video streams from separate cameras. All devices can work as validators and perform the PoENF consensus protocol. Dell Optiplex 760 desktop functions as edge server, and five desktops are configured as sites in our private Swarm network. To initiate comparative evaluation between EconLedger and existing blockchain benchmarks, test cases are also conducted on Ethereum [44] and Tendermint [45] networks. In our private Ethereum network setup, six miners are deployed on six separate desktops. Tendermint runs on a test network with 20 validators, and each validator is hosted on a RPi device.

Table 2. Configuration of experimental nodes.

Device	Dell Optiplex-7010	Dell Optiplex 760	Raspberry Pi 4 Model B
CPU	Intel Core TM i5-3470 (4 cores), 3.2 GHz	Intel Core TM E8400 (2 cores), 3 GHz	Broadcom ARM Cortex A72 (ARMv8), 1.5 GHz
Memory	8 GB DDR3	4 GB DDR3	4 GB SDRAM
Storage	350 G HHD	250 G HHD	64 GB (microSD)
OS	Ubuntu 16.04	Ubuntu 16.04	Raspbian (Jessie)

5.2. Performance Evaluation

In order to evaluate the performance of the running EconLedger under an IoVT-based edge network environment, a set of experiments is conducted by executing multiple complete epoch cycles of PoENF consensus protocol within a dynasty. The computation costs by message encryption and decryption are not considered during the test. As Krum in PoENF requires $n \geq 2f + 3$ and voting-based chain finality depends on a majority condition that requires $n \geq 3f + 1$, the minimum PoENF committee size is five members such that any $K \geq 5$ can meet both security requirements. Given 60 s per sliding window used in ENF fluctuations extraction, we let the ENF proof vector size $d = 60$. We conducted 100 Monte Carlo test runs for each test scenario and used the average of results for evaluation.

5.2.1. Network Latency

Figure 4 presents the network latency for EconLedger with respect to completing an entire epoch round of PoENF consensus protocol given the number of validators varying from 5 to 20. For each test point, we let all validators perform tasks simultaneously and waited until the bundle of tasks is finished. The latency includes the round trip time (RTT) and service processing time on the remote host. Broadcasting an ENF tx needs $\mathcal{O}(K)$ communication complexity and $K \times \mathcal{O}(1)$ computation complexity for verification. Thus, the total complexity is $\mathcal{O}(K)$ such that latency of ENF collection T_{ec} is linearly scale to committee size K. Chain finality requires all validators to broadcast their vote among committee members so that it has the same complexity as ENF collection. Thus, the delay of chain finality T_{cf} is almost linear scale to K.

The green line in Figure 4 shows the latency of block proposal T_{bp}, which indicates how long a proposed block could be accepted by all validators in the PoENF committee. The communication complexity of block proposal is $\mathcal{O}(K)$, which is similar to ENF collection and chain finality. However, during block generation and verification processes, PoENF algorithm requires a validator using Equation (1) to compute ENF scores based on collected proofs from others, and it has computation complexity of $\mathcal{O}(K^2 d)$. As a result, the total complexity of block proposal is $\mathcal{O}(K^2 d + K)$. In general, ENF samples size d is a small value such as 60, and it has less effect on computation cost than K does. Thus, T_{bp} is almost linearly scaled relative to $\mathcal{O}(K^2)$. The total latency shows that EconLedger takes about 2 s to finish an epoch cycle of PoENF committee consensus $(T_{ec} + T_{bp} + T_{cf})$ given the committee size $K = 20$.

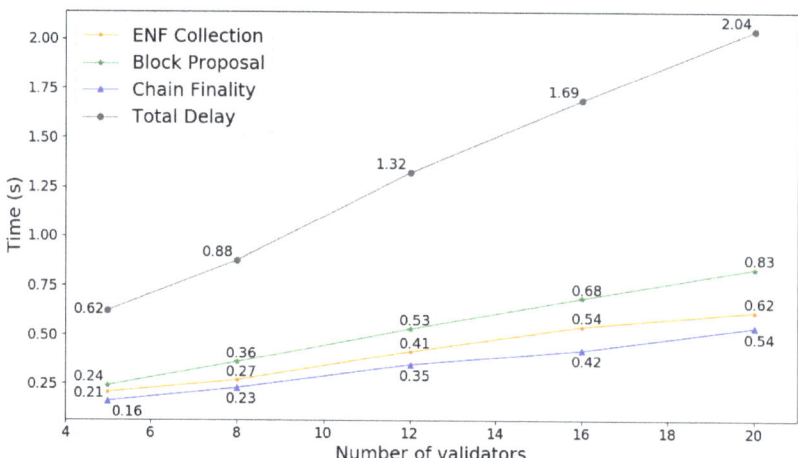

Figure 4. Latency for an epoch cycle of PoENF consensus with different committee size.

5.2.2. Computation Overhead

Figure 5 shows service processing time of key procedures in PoENF consensus given different platform benchmarks. As verifying a tx or vote only involves $\mathcal{O}(1)$ computation complexity, the service processing time is almost stable (about 2 ms for tx verification and 50 ms for vote verification) on all benchmarks. Compared to tx verification, which simply checks the validity of a tx then buffers it into system memory, vote verification involves more computation on resolving forks and database operations to store valid votes. Thus, vote verification incurs more latency than tx verification does. As the most computing intensive stages, both block mining and verifying rely on procedures in PoENF consensus algorithm with the computation complexity of $\mathcal{O}(K^2 d)$. Therefore, the computation cost on all devices dramatically increases as K is scaled up. Given different computation capacity of benchmarks, RPi-4 needs 2.5× processing time than Desktop does.

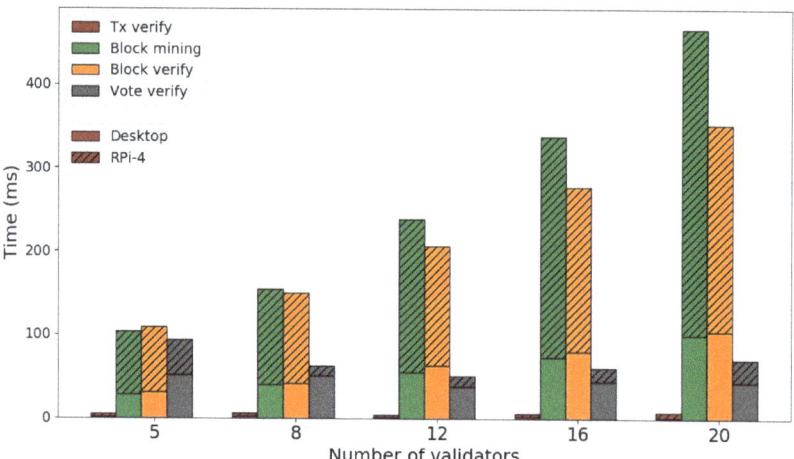

Figure 5. Computation overhead for stages of running PoENF consensus on host. Comparative evaluations on platform benchmark.

5.2.3. Data Throughput

In order to evaluate overhead of running EconLedger on communication channel, we considered volumes of message propagation and data throughput during key steps of the PoENF consensus protocol. Figure 6 demonstrates data transmission for different stages of PoENF consensus with varying committee size. In our EconLedger prototype, each ENF transaction has fixed size d_{tx} = 430 Bytes, and a vote has fixed size d_{vt} = 589 Bytes. Given the total communication complexity of $\mathcal{O}(K^2)$ in both ENF collection and chain finality, data transmission of ENF collection is $\mathcal{D}_{ec} = d_{tx} \times K^2$, and the data transmission of chain finality is $\mathcal{D}_{cf} = d_{vt} \times K^2$. Thus, communication overheads incurred by ENF collection and chain finality are linearly scaled to K^2.

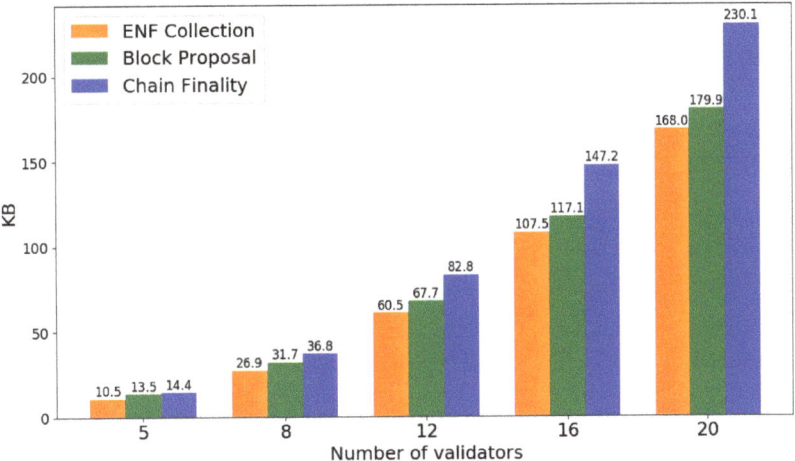

Figure 6. Communication overhead of running an epoch round of PoENF consensus. Comparative evaluations on different committee size.

Each block has a fixed header d_{head} = 613 Bytes along with a transactions list with size $d_{txs} = d_{tx} \times K$, and we can obtain block size $d_B = d_{head} + d_{txs}K$. Therefore, the block size d_B is linearly scaled to K. Assuming an ideal case that only one valid block is proposed during each epoch cycle, data transmission of block proposal is $\mathcal{D}_{bp} = d_B \times K = d_{head}K + d_{txs}K^2$ such that communication overhead is almost scaled to K^2. On the other hand, for the worst case that every validator proposed a candidate block such that $\mathcal{D}_{bp} = d_B \times K^2$, huge communication cost scaling up K can be introduced.

The data throughput could be specified as $Th = \frac{\mathcal{D}_{ec}+\mathcal{D}_{bp}+\mathcal{D}_{cf}}{\mathcal{T}_{ec}+\mathcal{T}_{bp}+\mathcal{T}_{cf}}$ (KB/s), where KB/s means KBytes per second. With variant committee sizes, the corresponding block size and data throughput are calculated as shown in Table 3. Given a fixed ENF transaction size, increasing the committee size allows committing more ENF proofs and, therefore, reach a higher data throughput at the cost of latency. In the test case of $K = 20$, EconLedger implies a theoretical maximum data rate of 283 KB/s, which can meet bandwidth conditions in a majority of LAN-based IoVT systems.

Table 3. Data throughputs vs. committee sizes.

Committee Size	5	8	12	16	20
Block Size (KB)	2.7	3.9	5.6	7.3	8.9
Throughput (KB/s)	61.9	108.3	159.8	220.1	283.3

5.3. Comparative Evaluation

As a key parameter in blockchain network, transaction tx committed time indicates how long a tx can be finalized in a new block on distributed ledger, and it is closely related to block confirmation time in consensus protocol. Given different blockchain benchmarks, we evaluate the end-to-end time latency of committing transactions along with other key performance metrics. For Ethereum, we used smart contract to record transactions on blockchain. For the Tendermint network, we used the built-in kvstore as an ABCI (Application BlockChain Interface) app to save transactions.

We conducted 50 Monte Carlo test runs, where a node sends a 1KB tx per second (TPS) to the blockchain network and waits until tx has been confirmed on the distributed ledger. Figure 7 shows the distributions of time delay for committing transactions given different blockchain networks. Each green bar indicates standard deviation with a mean represented by a red dot. The gray line shows the entire data range, and the black star is the median. Tendermint uses a BFT consensus protocol to achieve high efficiency; therefore, the mean of tx committed time is about 3 s given one voting round per second. Unlike Tendermint, Ethereum relies on probabilistic PoW consensus, which has variable block confirmation times. Thus, tx committed time in the Ethereum network varies with largest standard deviation. To guarantee synchronous epoch rounds for PoENF consensus, we set T_Δ conservatively to 2 s based on the maximum time to ensure txs and blocks propagation in a P2P consensus network, including 20 validators. Hence, the range of latency in EconLedger is smaller than Ethereum, and tx committed time is almost stable (about 5.5 s).

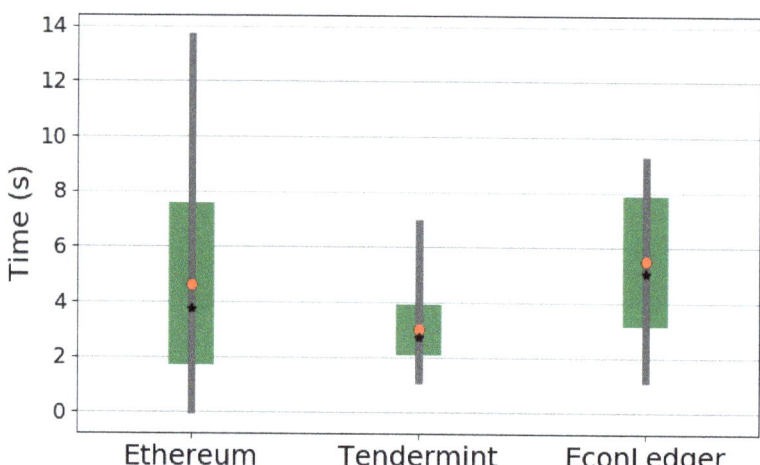

Figure 7. Time latency for committing transactions. Comparative evaluations on different blockchain networks.

Table 4 provides a comprehensive performance of committing transactions on different blockchain networks regarding several key performance matrices. Given the above tx committed time, which uses the mean in Figure 7, the tx rate tx/s is evaluated by calculating how many tx can be processed per second in the blockchain network. The Ethereum block size is bounded by how many units of gas can be spent per block, which is known as the block gas limit [46]. Currently, the maximum block size is around 12,000,000 gas (accessed at 20 July 2020), and the base cost of any transaction is about 21,000; thus, each block in Ethereum can include about 571 transactions. In our private Ethereum network, we can obtain the tx rate as $(571.4/4.6) \approx 124$ tx/s. Tendermint and EconLedger both use fixed 1MB block. Given 1 KB per transaction, a block in Tendermint can store a maximum of 1000 transactions; thus, the tx rate is about $(1000/2.9) \approx 344$ tx/s. For EconLedger, each transaction is about 430 bytes such that a block can record the maximum of 2400 transactions, then it achieves higher tx rate at around $(2400/5.5) \approx 436$ tx/s.

Table 4. Comparative evaluation of launching transactions on different blockchain platforms.

	Ethereum	Tendermint	EconLedger
tx committed time (s)	4.6	3.0	5.5
tx rate (tx/s)	124	334	436
CPU usage (%)	103	38.6	15.2
Memory usage (MB)	1200	72	80

In order to evaluate resource consumption by running blockchain benchmarks, we used the "top" command to monitor system performance of machines. We considered CPU and memory usage on Desktop (Ethereum miner) and Rpi (Tendermint and EconLedger validator). Due to computation intensive PoW algorithm, the mining process of Ethereum almost occupies full CPU capacity and consumes about 1.2 GB memory. Therefore, such a huge computation cost prevents resource constrained edge devices mining in Ethereum network. Unlike Ethereum, Tendermint and EconLedger use lightweight consensus algorithms to achieve efficiency in CPU and memory usage such that they are both suitable for deploying validators on edge devices. EconLedger almost has the same amount of memory usage as Tendermint in terms of running time. However, EconLedger has the higher tx committed time than Tendermint does, and it only needs 40% of the computation resource that Tendermint does.

5.4. Performance and Security Analysis

This section analyzes performance improvements given by the above experimental results and highlights the advantages of EconLedger compared with existing consensus protocols. Then, we evaluate security guarantees regarding committee selection and consensus algorithm. Finally, we list possible attacks and explain how EconLedge can prevent or mitigate these potential risks.

5.4.1. Performance Improvements

Given the above numerical results in terms of processing time and running time resource usages, our PoENF consensus is more computationally efficient than the PoW-based methods. Such a lightweight property of PoENF is promising for reducing energy consumption in mining processes and can lower demands on system capability for participants. Thus, resource-limited IoVT devices can directly work as validators (miners) rather than depending on support from an intermediate consensus layer by outsourcing mining tasks on fog networks or cloud servers. Compared with these hardware dependent solutions, such as REM based on Intel SGX and PoR requiring large local storage, our PoENF consensus relies on a platform independent algorithm to extract ENF-containing multimedia signals from recordings as ENF proofs. Therefore, it is promising to address heterogeneity issues as we integrate blockchain technology with IoVT systems that include multiple non-standard platforms.

EconLedger achieves communication efficiency by executing consensus protocol within a random selected PoENF committee. Such a small scale consensus network imposes low levels of data transfer overhead on IoVT systems at the network of edge, which has limited bandwidth. In addition, communication complexity for each validator is linearly scaled to PoENF committee size, as shown in Figure 6. Thus, limited data transmission also means lower energy consumption on devices during communication handling tasks. Unlike non-scalable BFT-based solutions that rely on a pre-fixed set of validators, EconLedger aims to improve scalability by requiring a randomly elected consensus committee to delegate other nodes of the network. As a tradeoff, EconLedger is actually a partially decentralized blockchain network.

In EconLedger, raw data are saved into off-chain storage deployed on a DDB network, while only references of data are encapsulated into transactions that are finalized on distributed ledger (on-chain storage). As a reference is a fixed length of hash value disregarding format or size of the source data, such light transactions can be used to verified complicated data in use over IoVT systems, such as multimedia recordings, contextual information, and trained models, etc. Moreover, each tx has fixed and small size such that a block can record more txs. As a result, the txs rate increased given that the block confirmation time is stable.

5.4.2. Committee Randomness Security

We assumed that an adversary has limited capacity such that he/she is subject to the usual cryptographic hardness assumptions and honest nodes never share their keys with each other or disclose the input string x of the VRF function before the end of randomness generation. Therefore, members of a new committee could be completely random owing to the unpredictability property of the VRF-based randomness string generation. In addition, given the assumption that an adversary can only control up to f byzantine validators, the chain finality achieves safety by making agreements on checkpoints if current PoENF committee has no less than $2f+1$ honest members. Therefore, the adversary has at most $m=1/4$ chance per round to control the checkpoint voting process. As a result, the probability that an adversary controls n consecutive checkpoint is upper-bounded by $P[X \geq n] = \frac{1}{4^n} < 10^{-\lambda}$. For $\lambda = 6$, the adversary will control at most ten consecutive chain finality runs.

5.4.3. PoENF Consensus Security

Unlike PoW and PoS consensus protocols that are vulnerable to mining centralization, whether a validator can become winner in current PoENF block proposal round depends on its ENF score rather than its controlled computation power or cryptocurrency stakes. Thus, an adversary cannot control the mining process by increasing investments on computation resource or owned coins. Moreover, the Krum rule adopted in ENF score calculation chooses $n-f-2$ closet ENF proofs and precludes the $f-1$ Byzantine proofs that are far away. Thus, all honest validators can output the same minimum ENF score as long as $n \geq 2f+3$, and our PoENF can prevent against ENF proof positioning attacks.

In PoENF consensus, all honest validators only accept valid blocks generated in the current epoch round; thus, *correctness (validity)* is ensured. In addition, PoENF achieves *consistency (agreement)* by requiring all honest validators to update their local chain head according to chain extension policies. At the end of a PoENF block proposal round, every honest validator should either accept valid transactions that are saved into a confirmed block as the local chain header or reject all transactions by extending an empty block on local chain header. Such a *liveness (termination)* property ensures that all valid ENF transactions are processed within the block generation round. Furthermore, voting-based chain finality can guarantee *safety*, which requires all honest validators to form a same total order of finalized blocks appended on the global unique main chain.

5.4.4. Analysis of Possible attacks

1. *Double spending attacks*: In a double spending scenario, an adversary attempts to revert a transaction that has been finalized on the distributed ledger. In Econledger, a voting-based chain finality mechanism ensures the total order and persistence of data recorded on the distributed ledger. Thus, once a transaction is finalized in the checkpoint block, all other honest validators will work on the finalized main chain and disregard any double spending transactions from attackers.

2. *Free-riding attacks*: There is a possibility of free-riding attacks that some lazy nodes only gain benefits by using the security service without fulfilling their responsibilities in the EconLedger network, such as forwarding messages or submitting ENF proofs. The

punishment strategies can prevent against free-riding attacks by reducing credit stake of dishonest nodes or even isolating them from the entire network.

3. *Selfish-mining attacks*: In a selfish-mining attack, the adversary tries to withhold blocks and release them strategically to reduce chain growth and increase the relative ratio of his proposed blocks. In PoENF consensus, only valid blocks generated in the current round can be accepted by honest validators, while those outdated blocks are discarded. Moreover, withholding blocks is a type of misbehavior in PoENF, and it decreases both profits and credit of a dishonest node. Therefore, selfish-mining is unprofitable for rational validators according to reward and punishment strategies.

4. *ENF-proof replay attacks*: The adversary can launch replay attacks by sending duplicated ENF proofs. As ENF fluctuations of power grid vary as time changes, the duplicate ENF proofs generally output large ENF scores. As a result, these Byzantine validators have marginal chances to propose valid blocks. Furthermore, ENF-proof replay attacks can be detected by analyzing ENF proofs on EconLedger. Thus, identifying misbehavior and isolating suspicious nodes can improve system robustness, while we leave ENF-based detection topics to future work.

6. Conclusions

This paper presents EconLedger, a lightweight and secure-by-design distributed ledger to enhance trust and security properties for smart IoVT systems at the edge. The EconLedger combines an efficient PoENF consensus mechanism with a deterministic voting-based chain finality in order to achieve safety and liveness. By using on-chain ledger and DDB enabled off-chain storage, the EconLedger network reduces storage overheads on validators and guarantees security and resilience of data sharing in a distributed IoVT network. The experimental results based on a prototype demonstrate that it achieves higher computation efficiency and tx throughput than benchmarks.

The experimental results on the prototype are encouraging, but there still are open issues to solve before developing a practical solution in real-world video surveillance systems. Using ENF signals for proof of work in consensus process is creative, however, whether ENF variation extracted from multimedia is reliable given attacks on ENF recordings such as synchronizing ENF and injecting into raw video/audio data or colluding among adversaries by sharing ENF data, is still an open question. Thus, our ongoing efforts include validating the proposed architecture in a real-world video streamscontext, simulating attack scenarios such as using AI enabled methods to generate fake ENF recordings, and ensuring overall efficiency and security.

In addition, validators in EconLedger system cannot directly obtain cryptocurrency rewards though PoENF consensus, but they can gain benefits from transaction fees. As a punishment strategy, slashing security deposits can increase financial cost if the adversary uses sybil nodes to disturb consensus protocol. However, there are open questions on the incentive mechanism. Our future work will use game theory to evaluate how incentive mechanisms can enhance system robustness and security.

Moreover, our EconLedger solution aims to provide a lightweight and security distributed ledger under a small-scale IoVT network, such as a campus. However, it still requires more investigation on how to apply EconLedger at a large-scale application scenario, such as smart cities or smart grids. Another future investigation for our team is designing scalable blockchain infrastructure that relies on a hierarchical framework in order to federate multiple privately distributed ledgers.

Author Contributions: Conceptualization, R.X., D.N., and Y.C.; methodology, R.X. and D.N.; software, R.X. and D.N.; validation, R.X., D.N., and Y.C.; formal analysis, R.X., D.N., and Y.C.; investigation, R.X.; resources, R.X. and D.N.; data curation, R.X.; writing—original draft preparation, R.X. and D.N.; writing—review and editing, R.X. and Y.C.; visualization, R.X.; supervision, Y.C.; project Administration, Y.C. All authors have read and agreed to the published version of the manuscript.

Funding: This research was funded by National Science Foundation, grant number CNS-2039342 and United States Air Force Office of Scientific Research, grant number FA9550-21-1-0229.

Data Availability Statement: Not Applicable, the study does not report any data.

Acknowledgments: This work is supported by the U.S. National Science Foundation (NSF) via grant CNS-2039342 and the U.S. Air Force Office of Scientific Research (AFOSR) Dynamic Data and Information Processing Program (DDIP) via grant FA9550-21-1-0229. The views and conclusions contained herein are those of the authors and should not be interpreted as necessarily representing the official policies or endorsements, either expressed or implied, of the U.S. Air Force.

Conflicts of Interest: The authors declare no conflict of interest.

Abbreviations

Abbreviations

The following abbreviations are used in this manuscript:

ABCI	Application BlockChain Interface;
BFT	Byzantine Fault Tolerant;
CCD	Charge-coupled Device;
COMS	Complementary Metal Oxide Semiconductor;
CPU	Central Processing Unit;
DAG	Directed Acyclic Graph;
DDB	Decentralized Database;
DHT	Distributed Hash Table;
DLT	Distributed Ledger Technology;
DPA	Distributed Pre-image Archive;
ENF	Electrical Network Frequency;
IoVT	Internet of Video Things;
LAN	Local Area Network;
M2M	Machine-to-Machine;
P2P	Peer-to-Peer;
PBFT	Practical Byzantine Fault Tolerant;
PKI	Public Key Infrastructure;
PoENF	Proof-of-ENF;
PoET	Proof-of-Elapsed Time;
PoR	Proofs-of-Retrievability;
PoS	Proof-of-Stake;
PoUW	Proof-of-Useful-Work;
PoW	Proof-of-Work;
PoX	Proof-of-X-concept;
PVSS	Publicly Verifiable Secret Sharing;
REM	Resource Efficient Mining;
RSA	Rivest–Shamir–Adleman;
RTT	Round Trip Time;
RPi	Raspberry Pi;
SGX	Software Guard Extensions;
SMR	State Machine Replication;
TPS	Transactions per Second;
VRF	Verifiable Random Function.

References

1. Xu, R.; Nikouei, S.Y.; Chen, Y.; Polunchenko, A.; Song, S.; Deng, C.; Faughnan, T.R. Real-time human objects tracking for smart surveillance at the edge. In Proceedings of the 2018 IEEE International Conference on Communications (ICC), Kansas City, MO, USA, 20–24 May 2018; pp. 1–6.

2. Nikouei, S.Y.; Xu, R.; Nagothu, D.; Chen, Y.; Aved, A.; Blasch, E. Real-time index authentication for event-oriented surveillance video query using blockchain. In Proceedings of the 2018 IEEE International Smart Cities Conference (ISC2), Kansas City, MO, USA, 16–19 September 2018; pp. 1–8.
3. Nikouei, S.Y.; Chen, Y.; Aved, A.; Blasch, E.; Faughnan, T.R. I-safe: Instant suspicious activity identification at the edge using fuzzy decision making. In Proceedings of the 4th ACM/IEEE Symposium on Edge Computing, Washington, DC, USA, 7–9 November 2019; pp. 101–112.
4. Ali, M.S.; Vecchio, M.; Pincheira, M.; Dolui, K.; Antonelli, F.; Rehmani, M.H. Applications of blockchains in the Internet of Things: A comprehensive survey. *IEEE Commun. Surv. Tutor.* **2018**, *21*, 1676–1717. [CrossRef]
5. Nikouei, S.Y.; Chen, Y.; Faughnan, T.R. Smart surveillance as an edge service for real-time human detection and tracking. In Proceedings of the 2018 IEEE/ACM Symposium on Edge Computing (SEC), Seattle, WA, USA, 25–27 October 2018; pp. 336–337.
6. Zhou, Q.; Huang, H.; Zheng, Z.; Bian, J. Solutions to scalability of blockchain: A survey. *IEEE Access* **2020**, *8*, 16440–16455. [CrossRef]
7. Nakamoto, S. *Bitcoin: A Peer-to-Peer Electronic Cash System*. Technical Report, Manubot. 2019. Available online: https://bitcoin.org/bitcoin.pdf (accessed on 21 September 2021).
8. Miller, A.; Juels, A.; Shi, E.; Parno, B.; Katz, J. Permacoin: Repurposing bitcoin work for data preservation. In Proceedings of the 2014 IEEE Symposium on Security and Privacy, Berkeley, CA, USA, 18–21 May 2014; pp. 475–490.
9. Castro, M.; Liskov, B. Practical Byzantine fault tolerance. *OSDI* **1999**, *99*, 173–186.
10. Hajj-Ahmad, A.; Garg, R.; Wu, M. ENF-based region-of-recording identification for media signals. *IEEE Trans. Inf. Forensics Secur.* **2015**, *10*, 1125–1136. [CrossRef]
11. Nagothu, D.; Chen, Y.; Blasch, E.; Aved, A.; Zhu, S. Detecting malicious false frame injection attacks on surveillance systems at the edge using electrical network frequency signals. *Sensors* **2019**, *19*, 2424. [CrossRef] [PubMed]
12. Swarm. Available online: https://ethersphere.github.io/swarm-home/ (accessed on 2 January 2020)
13. Bollen, M.H.; Gu, I.Y. *Signal Processing of Power Quality Disturbances*; John Wiley & Sons: Hoboken, NJ, USA, 2006; Volume 30.
14. Grigoras, C. Applications of ENF analysis in forensic authentication of digital audio and video recordings. *J. Audio Eng. Soc.* **2009**, *57*, 643–661.
15. Hajj-Ahmad, A.; Garg, R.; Wu, M. Instantaneous frequency estimation and localization for ENF signals. In Proceedings of The 2012 Asia Pacific Signal and Information Processing Association Annual Summit and Conference, Lanzhou, China, 18–21 November 2012; pp. 1–10.
16. Nagothu, D.; Chen, Y.; Aved, A.; Blasch, E. Authenticating Video Feeds using Electric Network Frequency Estimation at the Edge. *EAI Endorsed Trans. Secur. Saf.* **2021**, *2018*, 168648.
17. Buterin, V. *A Next-Generation Smart Contract and Decentralized Application Platform*. White Paper. 2014. Available online: https://blockchainlab.com/pdf/Ethereum_white_papera_next_generation_smart_contract_and_decentralized_application_platform-vitalik-buterin.pdf (accessed on 21 September 2021).
18. Juels, A.; Kaliski, B.S., Jr. PORs: Proofs of retrievability for large files. In Proceedings of the 14th ACM Conference on Computer and Communications Security, New York, NY, USA, 29 October 2007; pp. 584–597.
19. Zhang, F.; Eyal, I.; Escriva, R.; Juels, A.; Van Renesse, R. REM: Resource-Efficient Mining for Blockchains. In Proceedings of the 26th USENIX Security Symposium (USENIX Security 17), Vancouver, BC, Canada, 16–18 August 2017; pp. 1427–1444.
20. Sawtooth Documentation. Available online: https://sawtooth.hyperledger.org/docs/core/releases/latest/introduction.html (accessed on 2 January 2020).
21. King, S.; Nadal, S. *Ppcoin: Peer-to-Peer Crypto-Currency with Proof-of-Stake*; Self-Published Paper; 2012; Volume 19. Available online: https://people.cs.georgetown.edu/~clay/classes/fall2017/835/papers/peercoin-paper.pdf (accessed on 21 September 2021).
22. Lamport, L.; Shostak, R.; Pease, M. The Byzantine generals problem. *ACM Trans. Program. Lang. Syst. (TOPLAS)* **1982**, *4*, 382–401. [CrossRef]
23. Hyperledger Fabric. Available online: http://https://github.com/hyperledger/fabric (accessed on 2 January 2020).
24. Buterin, V.; Griffith, V. Casper the friendly finality gadget. *arXiv* **2017**, arXiv:1710.09437.
25. Bao, Z.; Shi, W.; He, D.; Chood, K.K.R. IoTChain: A three-tier blockchain-based IoT security architecture. *arXiv* **2018**, arXiv:1806.02008.
26. Tuli, S.; Mahmud, R.; Tuli, S.; Buyya, R. Fogbus: A blockchain-based lightweight framework for edge and fog computing. *J. Syst. Softw.* **2019**, *154*, 22–36. [CrossRef]
27. Sagirlar, G.; Carminati, B.; Ferrari, E.; Sheehan, J.D.; Ragnoli, E. Hybrid-iot: Hybrid blockchain architecture for internet of things-pow sub-blockchains. In Proceedings of the 2018 IEEE International Conference on Internet of Things (iThings) and IEEE Green Computing and Communications (GreenCom) and IEEE Cyber, Physical and Social Computing (CPSCom) and IEEE Smart Data (SmartData), Halifax, NS, Canada, 30 July–3 August 2018; pp. 1007–1016.
28. IOTA Foundation. IOTA Data Marketplace. Available online: https://data.iota.org (accessed on 2 January 2020).
29. Popov, S. The Tangle. 2018; p. 131. Available online: https://assets.ctfassets.net/r1dr6vzfxhev/2t4uxvsIqk0EUau6g2sw0g/45eae33637ca92f85dd9f4a3a218e1ec/iota1_4_3.pdf (accessed on 21 September 2021).
30. Xu, R.; Chen, Y.; Blasch, E.; Chen, G. Exploration of blockchain-enabled decentralized capability-based access control strategy for space situation awareness. *Opt. Eng.* **2019**, *58*, 041609. [CrossRef]

31. Xu, R.; Chen, Y.; Blasch, E.; Chen, G. Blendcac: A smart contract enabled decentralized capability-based access control mechanism for the iot. *Computers* **2018**, *7*, 39. [CrossRef]
32. Nagothu, D.; Schwell, J.; Chen, Y.; Blasch, E.; Zhu, S. A study on smart online frame forging attacks against video surveillance system. In *Sensors and Systems for Space Applications XII*; International Society for Optics and Photonics: 2019; Volume 11017, p. 110170L. Available online: https://arxiv.org/pdf/1903.03473.pdf (accessed on 21 September 2021).
33. Xu, R.; Nagothu, D.; Chen, Y. Decentralized video input authentication as an edge service for smart cities. *IEEE Consum. Electron. Mag.* **2021**. [CrossRef]
34. Lamport, L. Time, clocks, and the ordering of events in a distributed system. *Commun. ACM* **1978**, *21*, 558–565. [CrossRef]
35. Swarm Docs. Available online: https://swarm-guide.readthedocs.io/en/latest/introduction.html (accessed on 2 January 2020).
36. Maymounkov, P.; Mazieres, D. Kademlia: A peer-to-peer information system based on the xor metric. In *International Workshop on Peer-to-Peer Systems*; Springer: Berlin/Heidelberg, Germany, 2002; pp. 53–65.
37. Gilad, Y.; Hemo, R.; Micali, S.; Vlachos, G.; Zeldovich, N. Algorand: Scaling byzantine agreements for cryptocurrencies. In Proceedings of the 26th Symposium on Operating Systems Principles, Shanghai, China, 28–31 October 2017; pp. 51–68.
38. Stadler, M. Publicly verifiable secret sharing. In *International Conference on the Theory and Applications of Cryptographic Techniques*; Springer: Berlin/Heidelberg, Germany, 1996; pp. 190–199.
39. Blanchard, P.; Guerraoui, R.; Stainer, J. Machine learning with adversaries: Byzantine tolerant gradient descent. In Proceedings of the 31st International Conference on Neural Information Processing Systems, Long Beach, CA, USA, 4–9 December 2017; pp. 119–129.
40. Xu, R.; Chen, Y.; Blasch, E. Microchain: A Light Hierarchical Consensus Protocol for IoT Systems. In *Blockchain Applications in IoT Ecosystem*; Springer: Berlin/Heidelberg, Germany, 2021; pp. 129–149.
41. Flask: A Pyhon Microframework. Available online: http://flask.pocoo.org/ (accessed on 2 January 2020).
42. Pyca/Cryptography Documentation. Available online: http://pyca/cryptography (accessed on 2 January 2020).
43. SQLite. Available online: https://www.sqlite.org/index.html (accessed on 2 January 2020).
44. Ethereum Homestead Documentation. Available online: http://www.ethdocs.org/en/latest/index.html (accessed on 2 January 2020).
45. Kwon, J. Tendermint: Consensus without mining. *Draft Fall* **2014**, *1*, 11.
46. What's the Maximum Ethereum Block Size? Available online: https://ethgasstation.info/blog/ethereum-block-size/ (accessed on 2 January 2020).

 *future internet*

Article

Toward Blockchain-Enabled Supply Chain Anti-Counterfeiting and Traceability

Neo C. K. Yiu

Department of Computer Science, University of Oxford, Oxford OX1 3QD, UK;
neo-chungkit.yiu@kellogg.ox.ac.uk

Abstract: Existing product anti-counterfeiting and traceability solutions across today's internationally spanning supply chain networks are indeed developed and implemented with centralized system architecture relying on centralized authorities or intermediaries. Vulnerabilities of centralized product anti-counterfeiting solutions could possibly lead to system failure or susceptibility of malicious modifications performed on product records or various potential attacks to the system components by dishonest participant nodes traversing along the supply chain. Blockchain technology has progressed from simply being a use case of immutable ledger for cryptocurrency transactions, to a programmable interactive environment of developing decentralized and reliable applications addressing different use cases globally. Key areas of decentralization, fundamental system requirements, and feasible mechanisms of developing decentralized product anti-counterfeiting and traceability ecosystems utilizing blockchain technology are identified in this research, via a series of security analyses performed against solutions currently implemented in supply chain industry with centralized architecture. The decentralized solution will be a secure and immutable scientific data provenance tracking and management platform where provenance records, providing compelling properties on data integrity of luxurious goods, are recorded and verified automatically across the supply chain.

Keywords: blockchain; anti-counterfeiting; decentralization; product authenticity; end-to-end traceability; supply chain integrity; supply chain provenance; NFC-Enabled Anti-Counterfeiting System; Near-Field communication; Internet-of-Things

1. Introduction

The problem of counterfeit product trading, including luxurious goods or pharmaceutical products, has been one of the major challenges the supply chain industry has been facing in an innovation-driven global economy. The situation has been alarming with an exponential growth of counterfeit and pirated goods worldwide, for which it has also plagued companies with multinational supply chain networks.

The analytical study [1] published by the Organization for Economic Cooperation and Development (OECD) and European Union Intellectual Property Office (EUIPO) in 2019 regarding the global trading activities of counterfeit and pirated products has suggested the volume of international trade of counterfeit and pirated products increased from $250 billion in 2007, to up to $461 billion in 2013, representing approximately 2.5% of world imports. The imports of counterfeit and pirated products into EU amounted to nearly $116 billion, representing 5% of EU imports approximately. The results have been alarming according to the latest statistics published in 2016, suggested that the volume has already further amounted to as much as $509 billion representing 3.3% of world trade and 6.8% of imports from non-EU countries.

The battle against counterfeit product trading remains a significant challenge; it is simultaneously of significant and growing concern to the globalized economy not to mention all the affected industries, such as the markets included in [2] and innumerable branded product companies. Trading activities in counterfeit goods not only infringe on trademarks and copyrights, and negatively impact on sales and profits of industries, but also generate profits for organized crime at the expense of the affected companies and governments. Counterfeit trading activities also pose broader adverse effects to the economy, public health, safety, and security of the wider community. Counterfeit product trading activities are operated swiftly in the globalized economy, misusing free trade zones according to the authors of [3], taking advantage of many legitimate trade facilitation mechanisms and thriving in economies with weak governance standard and limited innovative options to combat product counterfeits. In response to the growing concern in product counterfeit, innovative, fully functional, integrable, and affordable product anti-counterfeiting solutions with traceability functionalities have been widely and urgently demanded. These solutions are expected to utilize cutting-edge technologies so as to ensure the provenance and traceability of genuine products throughout the supply chain counterfeiting, and these suggested solutions should be widely adopted regardless of industries, size of the companies, and its supply chain systems.

Given the growing concern in counterfeit trading activities, though there have already been a variety of innovative product anti-counterfeiting solutions introduced in supply chain industry, the main research question is as follows:

"Why would existing anti-counterfeiting and traceability systems benefit from decentralization enabled by blockchain technology to better combat the rampant counterfeiting attacks?"

The main question could further be addressed via answering the following sub-questions throughout the research:

1. Why can blockchain be utilized to decentralize the supply chain anti-counterfeiting and traceability system?
2. What are the security limitations and concerns of existing supply chain anti-counterfeiting and traceability systems?
3. What are the advantages introduced by decentralized supply chain anti-counterfeiting and traceability systems?

Given the main research question and the set of sub-questions derived from it, it is common to follow an organized way of exploring them stepwise in this research. An exploratory research method, as depicted in [4], will be adopted in this research in computing. Contrary to existing conceptual blockchain implementations applied in different industries with less explanation on why decentralization is needed for these use cases, this research is taking a rather different approach to have a thorough process involving a series of security analyses on an existing product anti-counterfeiting and traceability system—NFC-Enabled Anti-Counterfeiting System (NAS). The insightful findings from the security analyses will further elaborate why these solutions could benefit from decentralization, on which a decentralized version of these supply chain software solutions could be developed and implemented.

The fundamental ideas and use cases of blockchain technology, categorized into different phases, are explained in the research. The reasons why implementations of Blockchain 2.0 should be adopted to decentralize the existing product anti-counterfeiting and traceability systems are also elaborated, before stepping into the overview on one of the existing product anti-counterfeiting and traceability system—NAS. This research will focus on explaining the reasons why these existing supply chain software solutions, including NAS, would benefit from decentralization enabled by blockchain technology.

Based on the findings and opportunities identified from these analyses, a set of fundamental system requirements of developing a decentralized product anti-counterfeiting and traceability system is also defined. The decentralized solutions, such as the Decentralized NFC-Enabled Anti-Counterfeiting System (dNAS), are aimed at delivering a more secure and higher quality approach to verify authenticity and provenance of luxurious products, such as bottled wine. With the use of peer-to-peer blockchain networks and distributed storage technologies, it will be possible to eliminate an absurd amount of cost for on-chain storage and provide a much higher level of privacy, reliability, and quality of service compared with existing anti-counterfeiting and traceability solutions with centralized architecture. The decentralized version of the existing solutions, such as dNAS, could also define a framework and practice for different nodes along the supply chain to integrate the low-cost, real-time, and immutable blockchain technology into their daily supply chain workflows.

2. Background

Given the growing concern around the trading activities of counterfeit products, anti-counterfeiting solutions have been developed and implemented in the supply chain systems of different industries.

Related Work of Anti-Counterfeiting and Traceability Systems

The starting point, such as in [5], is always digitalizing the multi-node supply chain system under which the supply chain network will at least be operating with a Point-of-Sale (POS) system integrated with the states of a data storage system updated at different nodes along the supply chain. Wireless communication technologies, such as Radio-frequency Identification (RFID) or near-field communication (NFC), which is a subset of RFID, powering the Internet of Things (IoT) are mostly the existing solutions that the centralized architectures are currently based on. The tags with wireless communication capabilities are packaged on the packets of goods or the product itself for identification, anti-counterfeiting, and traceability purposes.

The RFID-based solutions in [6,7] and NFC-based solutions in [8,9] both require dedicated applications or authentication servers to integrate with these tags so as to perform writing, reading, and validating features on the data stored in these tags with the product identifier (PID). Its metadata are also stored in a database system of applications for the purpose of validation at later stage to respond to any potential counterfeiting attack at different supply chain nodes. The work in [10] utilized simple barcode technology, such as QR code, to retrieve the product identifiers by scanning the products and querying corresponding records throughout the dedicated backend database systems to retrieve metadata of specific PIDs so as to prove the authenticity with the PIDs validated and detailed product data retrieved, such as in [11] where this method was implemented for drug anti-counterfeiting.

For instance, with the wider adoption of these wireless communication and barcode technologies, the concept of RFID-enabled track-and-trace anti-counterfeiting was also proposed in [12,13], respectively, to combat counterfeiting activities. RFID-enabled anti-counterfeiting approaches are further analyzed and compared with potential implementation issues in [7,14]. The *EPCglobal* traceable system to support anti-counterfeiting is further extended in [15,16]. The authors of [17] suggested the use of a Wireless Sensor Network (WSN) and geospatial technologies, such as Geographic Information System (GIS), Global Positioning System (GPS), and Remote Sensing (RS), for mobile assets-tracking operations; these are among the technologies that have been applied in product traceability systems.

Nevertheless, many of the existing implementations for product anti-counterfeiting in the supply chain industry are indeed built with centralized architecture relying on a trusted server, database, and applications. These implementations are solely controlled and managed by the manufacturers or suppliers of the products, for coordinating and managing product authentication with participation contributed from different nodes along the supply chain of different industries.

3. Decentralizing with the Blockchain Technology

In this section, following the advantages introduced by the blockchain technology, the modern core blockchain concepts including those of blockchain 1.0 and blockchain 2.0 are explained. A variety of current blockchain implementations applied to different types of supply chain industries is also mentioned. The reasons why developing decentralized solutions based on existing systems of centralized architecture is a pragmatic way going forward to improve the challenging product counterfeiting in the wider supply chain industry are also given.

3.1. Blockchain at the Core

The decentralized architecture could bring more advantages to the existing centralized product anti-counterfeiting system, and an example could be decentralizing NAS utilizing blockchain technology. As described in the first Blockchain use case [18], Blockchain is a distributed ledger technology recording and sharing all the transactions that occur within a dedicated peer-to-peer network. It is essentially a decentralized timestamp service with a virtual machine to execute signed scripts that operates on signed data. It utilizes a distributed ledger to store scripts and data with mutual consensus reached among participating nodes running on the same blockchain network.

The blockchain network consists of multiple nodes that maintain a set of shared states and perform transactions updating the states which could be divided into ledger state, block state, and transaction state as depicted in Figure 1. Blockchain transactions, as described in [19], need to go through the mining process. The transactions must be validated by the majority or agreed fraction among the participating network nodes, depending on which consensus protocol is adopted, before being ordered and packaged into a timestamped block which is also known as *block signing*.

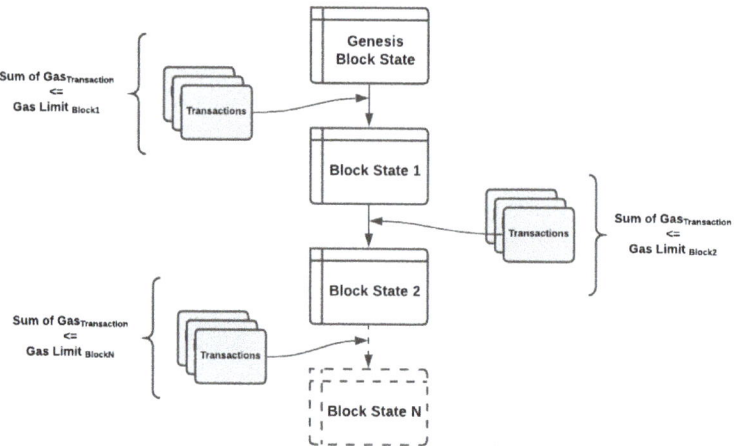

Figure 1. Block state with transactions flow.

The blockchain network can be generally categorized either as permissionless (public network) or permissioned network. The former is an open distributed ledger network, such as in [18,20], where any node can join the network and where any two peers can conduct transactions without any authentication performed by any central authority. The latter is a controlled distributed ledger (like in [21]) where the decision-making and validation process are kept to one organization or few organizations forming a consortium with or without the staking concept. In permissioned networks, the consortium administrator or certificate authority determines who can join the network as a validator node or listener node, if there is no logic of on-chain governance available. All nodes are authenticated in advance, and their identities are known to other nodes running on the same network and in the same consortium, at least to the administrator.

The general blockchain data structure is demonstrated in Figure 2. The first block is always referred as the *genesis block*, and a block consists of a header and a body. The block body contains the list of transactions. The number of transactions that can fit into a block is dependent on the block size (block gas limit) and the transaction size (gas spent per transaction). The block header, as discussed in [22], contains a wide variety of fields, including timestamp, Merkle root hash representing the hash value of every transaction in the block, and the hash pointer of the previous block header for which different blocks are "*chained*" to each other by putting this field of hash for every next block. There are more fields, such as the nonce which is the 32-bit field incremented until the equation is solved and difficulty which is needed for the *Proof-of-Work (PoW)* protocol. PoW is heavily linked with computation process known as *mining*, for which miners are the nodes to calculate the block header hash termed as "*solving the puzzle*". The differences between blockchains and databases are also explained in [23].

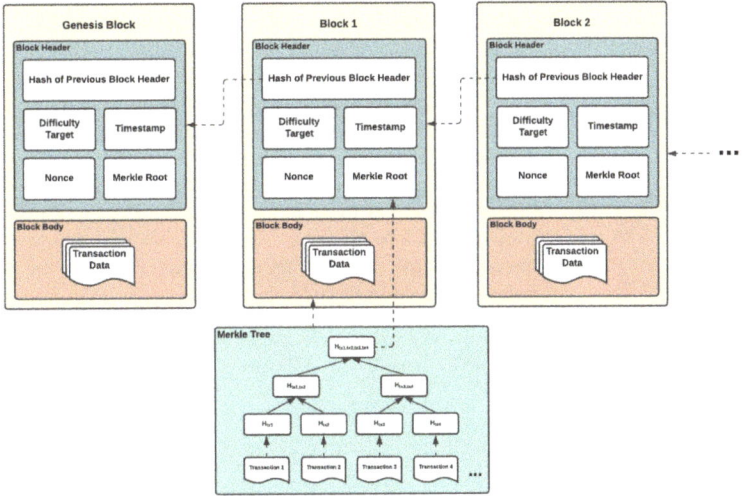

Figure 2. General blockchain data structure.

Based on the Proof-of-Work consensus protocol, the block is said to be mined if a miner finds its nonce such that the hash of block header is less than the difficulty target, based on the work in [18]. Modern blockchain is also characterized into four main aspects, apart from being utilized merely as the distributed ledger: the self-executing smart contract, immutability, cryptography, and consensus.

3.2. Starting from the Original Blockchain 1.0—The Bitcoin Network

Blockchain is often regarded as the underlying technology of Bitcoin [18]—peer-to-peer version of electronic cash, namely, the decentralized virtual currency, which does not require any existing currency institutions to circulate and is of fixed currency circulation. The Bitcoin network is indeed the first use case adopting blockchain technology. Bitcoin aimed at offering a purely peer-to-peer version of electronic cash which would allow online payments to be sent directly from one party to another without involving a financial institution. The main benefits of such a decentralized virtual currency system are the prevention of double spending, single point of control, and potential failure due to the reliance of trusted third parties and intermediaries. The Bitcoin network relies heavily on decentralized consensus and its cryptographic properties with use of digital signature instead, offering new transparency to finance industry, which have normally been of great security concerns on virtual currencies.

Blocks of the Bitcoin network are mined through a computationally-intensive process also known as the Proof-of-Work consensus protocol. The detailed process of PoW is depicted in [24], requiring significant computational resources to solve a cryptographic hash-based puzzle, and the solution could be worked out by trial-and-error based on the targeted difficulty set per block. The consensus must be reached before a new block could be created with respective transactions packed in the new block. As there are many miner nodes available on the open Bitcoin network, every miner on the network competes to generate a valid Proof-of-Work consensus for the block. It will take approximately 10 min on average with the current setting of the Bitcoin network for a miner to create a block successfully and receive the mining reward which has been halved on predefined milestone blocks (also known as the "*halving*" as explained in [25]) of the Bitcoin network. The Proof-of-Work adopted in the Bitcoin network would prevent the Sybil attackers from promoting a dishonest blockchain supporting their malicious agendas, offering a way for honest nodes to overcome Byzantine failures as well as accepting the next block on the canonical chain. This process is arguably the most difficult part of implementing a consensus protocol where many attack vectors would be focused on, for which a Byzantine failure (or fault) is a condition of a distributed network, where participating nodes may fail, and there is imperfect information on whether a specific node has failed.

There are also conditions for which a transaction in the Bitcoin network would be validated and so a successful state transition would then attain. For instance, (1) digital assets involved in the transaction of transfer operations should exist, (2) by enforcing asymmetric cryptography to produce signatures every node should only spend the coins they own and not those of others, and (3) every transaction should be supplied with enough values to the inputs field of every transaction by summing up all the Unspent Transaction Outputs (*UTXOs*) the sending blockchain nodes owned. The concept of *UTXO* is demonstrated in Figure 3.

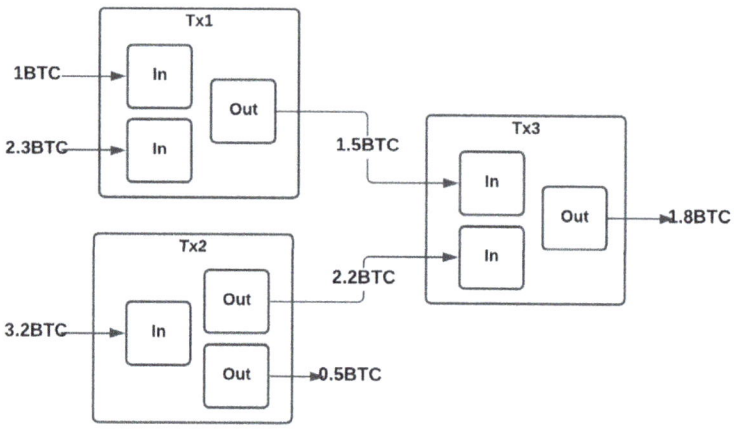

Figure 3. Concept of Unspent Transaction Outputs (UTXOs).

With the scripting ability of the Bitcoin network alongside its Proof-of-Work consensus algorithm requiring validation performed by participating nodes when the state-transitioned function is validated, the faulty transactions, such as the one sending the same fund twice, will receive an error and therefore be aborted. However, some malicious nodes could try to fork the chain and place a second transaction before the first requiring the calculation of upcoming blocks with the updated block headers, which would require the creation of a separate chain longer than the original chain to be the canonical one as nodes are programmed to settle on the chain with largest investment value which is the *canonical chain*. The authors of [26] suggested that the Bitcoin network could not actually solve the Byzantine Generals problem in general, as attackers could theoretically be computationally unlimited and dominate more than 51% share of the computation power. The overall mining hash rate of the network to perform double-spend operation faster than that on the canonical chain, also known as the 51% attack under which the analysis on the probability of solving n number of blocks consecutively faster than the canonical chain is demonstrated in Figure 4.

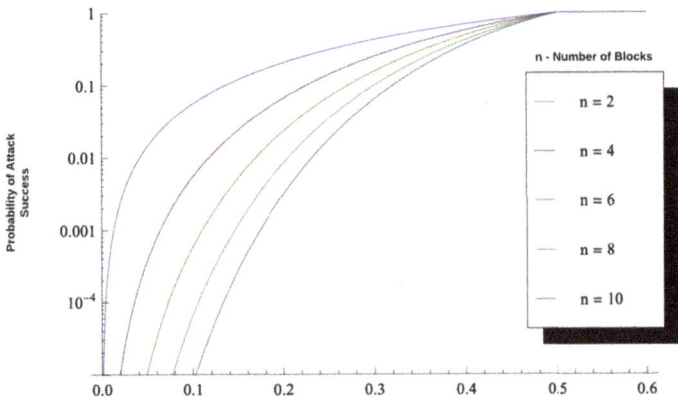

Figure 4. Analysis of hash rate-based Double-Spending.

There are research and development efforts performed based on the Bitcoin model in the field of decentralized electronic payment, such as Litecoin in [27]. These counterparts, being anything other than the original Bitcoin, were grouped as *"altcoins"*, in which some basically are hard-forked versions of Bitcoin, while others have their own underlying native blockchain network with their own consensus protocols, such as Ethereum. With the advent of increasingly more native blockchain networks with their own consensus protocols and proposed data structures to be supported, blockchain provides a way for untrusting parties on a peer-to-peer network to agree on contents of a vastly replicated database. The blockchain industry has been focused on exploring more use cases other than merely the decentralized electronic payment using blockchain technology. The development of Blockchain 1.0 has undoubtedly set the premise for new ideas around decentralized autonomous organization and provided a solid basis for the development of Blockchain 2.0 protocols.

3.3. Overview of Blockchain 2.0—The Programmable Blockchain

Given the fact that the Bitcoin network only offers basic scripting functionality, with the advent of the open-source Ethereum, which was published as in [28] back in 2014, Ethereum is no longer limited to transaction records, and is more effective and robust than its counterpart Bitcoin. The Ethereum blockchain network is a programmable blockchain that can perform any arbitrarily complex computation unlike those predefined operations performed in transactions of Bitcoin. Ethereum allows developers to create their own operations of any complexity in smart contracts, utilizing the Turing-completeness programming language and the flexibility brought by the smart contract enabling more possibilities to the blockchain. Ethereum has therefore often been dubbed as Web 3.0 due to the fact that the architecture of Ethereum opens up more ideas of general applications with transactions related to data processing and transfer of digital assets, not only the typical use cases, such as decentralized cryptocurrencies.

3.3.1. States and Accounts

To get started, Ethereum is an *account-based* blockchain, instead of the Unspent Transaction Outputs (UTXO)-based blockchain like Bitcoin network, under which all account states stored locally as a form of state data with the predefined data structure of *Merkle Patricia tree*. As described in [20], Ethereum blocks contain a list of transactions and the Merkle root hash of entire state tree on transactions which are packed in every block. Every node on the network stores two types of states: the overall blockchain state and the state data containing a list of accounts and their associated states.

There are also two general types of Ethereum accounts: (1) *externally owned accounts (EOAs)* with key pairs generated and assigned for signature-based validations or signing transactions on the network, and (2) *contract accounts*, which essentially act as autonomous agents containing code that only act once they receive a message that has initiated its functionalities to read from and write to internal storage of the deployed smart contracts. The authors of [28,29] also proved that contract accounts can send messages to its counterparts with embedded calls to methods of other deployed smart contracts, which is basically another type of account on Ethereum blockchain.

3.3.2. Smart Contracts and Ethereum Virtual Machine

Entities on any consensus mode of Ethereum network are able to write smart contracts with methods to define transaction formats, access permission of the methods, state conversion equations suggested in [30], and literally any self-defined rules applied to method declarations with examples demonstrated in [31].

Entities in the network could first write and deploy a smart contract to the network for its decentralized applications to interact with, via its dedicated node using the blockchain client. The smart contract source code, written in Solidity for instance, is then encoded into *Ethereum bytecode* by the Solidity compiler. The bytecode is added to the data field of a transaction and deployed to the transaction pool of the network queuing to be picked up for further processes. The typical workflow of the smart contract source code is described in Figure 5.

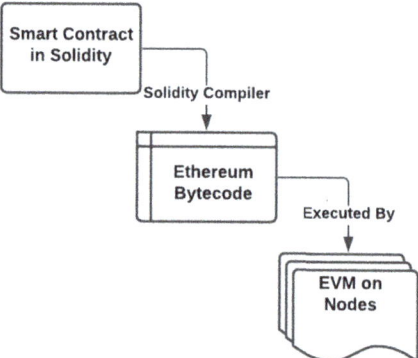

Figure 5. The workflow of Ethereum smart contract source.

As detailed in [32], the miners would then pick up the transaction via its node client, pack the validated transaction into a new block, and run the hexadecimal bytecode data of the transaction with its Ethereum Virtual Machine (EVM), which is the execution environment for running transaction code to reach a consensus. The work in [28] details a more complex set of instructions to be compiled in the form of smart contracts, to generate operation codes (e.g., *PUSH1 0 × 60 PUSH1 0 × 40 MSTORE*) and run based on the operation codes each time a specific method in the smart contract related to a specific transaction that is invoked following the rules set within the deployed smart contract. The smart contract could then have state transitioned on each miner's local persistent storage on state data, only if data of the transaction are executed by the EVM successfully. For miner nodes on the network to be able to validate state changes brought by the transaction with the deployed smart contracts, they would be required to run the data, which is the bytecode with operation codes, retrieved from the transaction, on its EVM to check if it is actually a valid state transition. This would impose a large computational redundancy on the network, but it is necessary in order to reach the decentralized consensus.

There is also an inherent constraint on the number of computational steps a transaction can perform, which is limited with a concept of *"Gas"*—essentially the cost incurred by totaling all the methods of a deployed smart contract executed in a submitted transaction. The cost calculation is based on the gas spending guideline of each functional pattern predefined in [28]. Nodes will need to specify a maximum amount of gas they are willing to pay per transaction, and if not explicitly stated, then the client implementation of the network (e.g., Go-Ethereum, Parity) would determine the amount automatically. The storage shall be as lean as possible as it could cost an absurd amount of resources. There are current technologies which could enable decentralized data storage, such as in [33], and therefore decentralized solutions could be built and not required to stick with central database architecture. Every block, created by miners after the consensus reached among nodes running on the network, has a block gas limit similar to the transaction gas limit. The gas limit is the upper bound on the amount of computation a transaction can perform on its workflow in the network, and that is why people think Ether as the crypto fuels of Ethereum due to a fact that gas cost will be paid in Ether.

3.4. Blockchain 1.0 Versus Blockchain 2.0 and Later Versions

Blockchain implementations have been phased based on their development and use cases described in [34,35], respectively. While Blockchain 1.0, with the representative example of the Bitcoin network [18], focused on the development of cryptocurrencies for peer-to-peer electronic payments, Blockchain 2.0, such as Ethereum explained in [20,28] as well as Hyperledger Fabric discussed in [21], enabled the concept of smart contract. Blockchain 2.0 has a more flexible data structure and functionality enabled by deployed smart contracts. Blockchain 3.0 focuses on developing ĐApps essentially having back-end logic patterns running on a decentralized peer-to-peer platform with a dedicated user interface of it. Blockchain 4.0 focuses more on matching blockchain technology and its implementations usable especially to those Industry 4.0 business demands, such as process automation, enterprise resource planning, and integration of different execution systems respectively.

Starting from Blockchain 1.0, the Bitcoin network [18] is indeed the first use case of blockchain technology, aimed at offering a peer-to-peer version of electronic cash. The Proof-of-Work consensus algorithm applied to Bitcoin network, its unique *UTXOs* model, and the potential hash rate-based double-spending in [36], which could still exist in Bitcoin network, are the major characteristics and topics discussed and advanced among Bitcoin or even the entire blockchain development community. Blockchain 2.0 moved on enabling programmable blockchains with Ethereum. A variety of underlying concepts of Ethereum blockchain, such as account-based, smart contracts, gas, and Ethereum Virtual Machine, as well as explaining how Ethereum, is designed and progressed to be different from, and even more capable than, Bitcoin network, are explained in [20,28], respectively, as representative examples.

There are indeed extensive differences in terms of implementation and usage in both phases. Figure 6 demonstrates the most crucial differences in different perspectives, followed by explanations on why modern decentralized solutions would be developed with frameworks and functionalities provided and enabled in Blockchain 2.0.

Properties	Blockchain 1.0 (Bitcoin)	Blockchain 2.0 (e.g. Ethereum) and Later Versions
Type of State	Only Two States (Successful/Unsuccessful)	Multiple States
Block Time	Long (in minutes)	Short (in seconds)
Block Data	Fixed Script	Diversified Scripts
Turing Completeness	No	Yes

Figure 6. Comparison of Blockchain 1.0 and Blockchain 2.0 (including later versions).

- *State*: In Bitcoin, there are only two states, not to mention UTXO, that could either be "*spent*" or "*unspent*", and so there is no contract or script to keep any internal state other than these two states. Ethereum enables more flexibility to create such contracts by utilizing the concepts of *externally owned accounts* and *contract accounts*. The multi-state can be defined given the functionalities of smart contracts. Once transactions are validated, packaged into their respective mined blocks, and appended to the blockchain, they are no longer allowed to be modified. If such modification on the specific blocks took place, it will invalidate every subsequent block intended to be appended to the blockchain.
- *Block Time*: The creation of a block in the Ethereum main net takes only 12 s (while block time could be specified in enterprise implementations of Ethereum), which is considerably faster than that in Bitcoin network taking nearly 10 min. Every transaction will then be validated and packed in a block quicker due to the faster block time in Ethereum but it would lead to decreased security, attributed by the faster block time, which has already been addressed by multiple block confirmations.
- *Storage*: In the Bitcoin network, only fixed scripts and data can be stored in a block, while self-defined scripts, in a form of smart contracts, can be executed with states stored on Ethereum blockchain implying that more methods are enabled. Many different applications can therefore be implemented on Ethereum or its Blockchain 2.0 counterparts.
- *Turing Completeness*: The script in the Bitcoin network does not support loops, so infinite loops can be prevented, while Ethereum provides more flexibility in script writing of its smart contract implementations and Turing completeness as it employs different methods to eradicate infinite loops.

While the open-source Ethereum blockchain has been planning a major *Ethereum 2.0* (Eth2) upgrade to its network to address scalability concerns, there has been an array of blockchain frameworks in Blockchain 2.0. These frameworks are available for developing decentralized solutions based on a concept of enterprise blockchain, such as Hyperledger Fabric depicted in [21] or Tendermint Core developed based on the system approaches covered in [37,38]. Like Ethereum, all of these have been seeking to prove that enhanced security, enhanced degree of decentralization, and enhanced scalability are not at odds.

3.5. Related Work of Blockchain Implementations

With the advancement of blockchain development in recent years, there have been some existing blockchain innovations developed in different domains and in combination with other emerging technologies to decentralize software systems with centralized architecture of different purposes. A blockchain-based digital certificate system and blockchain-enabled system for personal data protection were proposed in [39,40], respectively. Furthermore, the works in [41,42] have also given an overview of blockchain-based applications developed in different domains where a variety of examples of blockchain systems and use cases are developed, could be found in healthcare domain, as depicted in [43,44], focused on decentralizing health record management and storage.

Blockchain innovations have also been implemented across the supply chain industry, and some are specifically with use cases of decentralizing and improving product traceability and anti-counterfeiting aspects of the supply chain industry. The authors of [45] have proposed a concept of a blockchain system to enhance transparency, traceability, and process integrity of manufacturing supply chains, while an Ethereum-based fully decentralized traceability system for Agri-food supply chain management named AgriBlockIoT was developed in [46]. Furthermore, there is a wide range of blockchain innovations applied in supply chain industry, such as the novel blockchain-based product ownership management system in [47], a blockchain-based anti-counterfeiting system coupled with chemical signature for additive manufacturing described in [48], and an ontology-driven blockchain design for supply chain provenance in [49], as well as a temperature monitoring and anti-counterfeiting system approach for pharmaceutical supply chains as depicted in [50].

Some implementations are conceptualized and developed based on computation-extensive permissionless blockchain networks and consensus algorithms, aiming at full decentralization over scalability and stability of such decentralized systems developed. Instead of developing blockchain implementations based on conceptual design, decentralizing legacy anti-counterfeiting systems with centralized architecture already implemented in the industry, further with blockchain innovations integrated with, would be a more pragmatic way to start with so as to provide timely support to improve the snowballing situation of product counterfeits in supply chain industry.

4. The NFC-Enabled Anti-Counterfeiting System—NAS

One of the solutions to answer the growing concerns on product counterfeiting in different supply chain systems of wine industry is the *NFC-enabled Anti-Counterfeiting System (NAS)*. The NAS in [51] was developed and implemented back in 2014, aiming at providing an innovative and fully functional alternative, based on *Near-field communication technology* and cloud-based microservices architecture with centralized storage structure, solely hosted by any winemaker, to help improve the worsening situation of product counterfeiting especially for the wine industry. NAS with centralized data architectures, which are predominantly based on typical and familiar cloud-based client–server architecture style, is demonstrated in Figure 7.

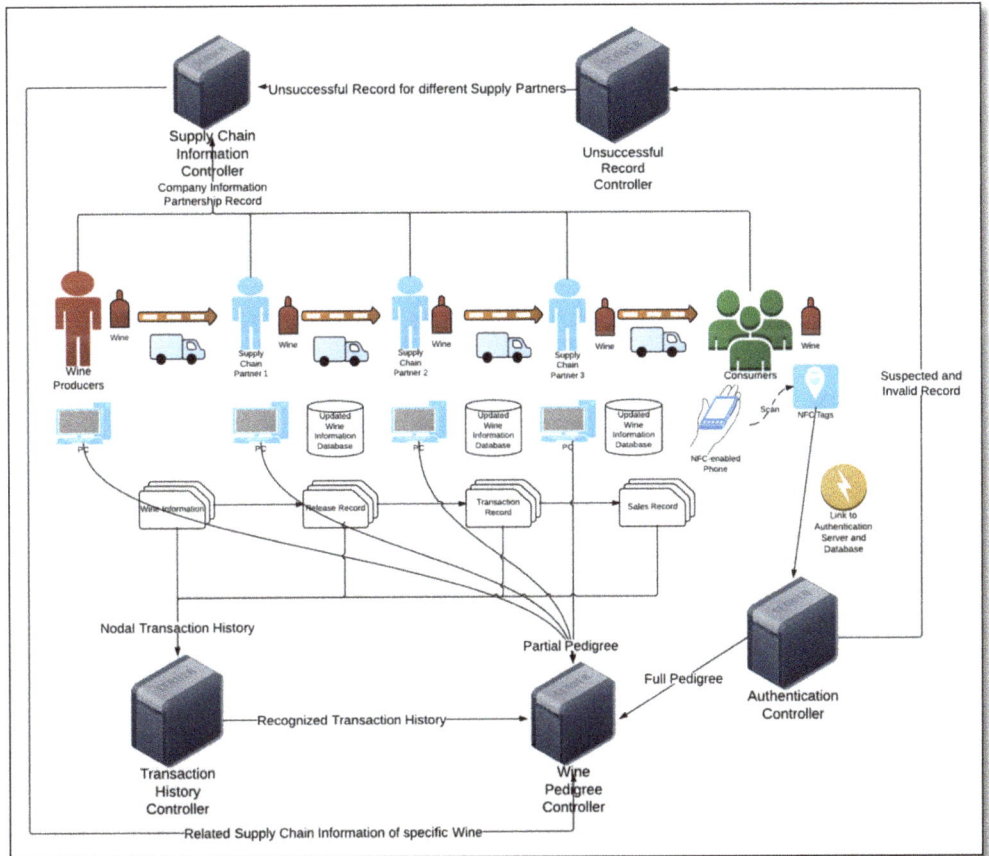

Figure 7. The System Architecture of NFC-Enabled Anti-Counterfeiting System (NAS). Source: Neo C. K. Yiu.

The whole NAS solution consists of five main components: (1) the back-end system with web-based database management user interface for wine data management performed by winemakers, for which management of data columns of specific wine products which are in their custody can be performed by winemakers; (2) an NFC-enabled mobile application, *ScanWINE*, for tag-reading purpose of wine products at retailer points for wine consumers or supply chain participants of the supply chain before accepting these wine products; (3) another NFC-enabled mobile application, *TagWINE*, performing tag writing purpose for wine at wine bottling stage by winemakers; (4) the NFC tags packaged on wine bottlenecks for those purposes and actions; and (5) any NFC-enabled smartphones or tablets of supply chain participants and wine consumers along the supply chain.

There are four major categories of data to be processed along the supply chain with NAS: (1) nodal transaction history data, (2) supply chain data, (3) wine pedigree data which is processed with its dedicated controllers based on their predefined schema and data models, and (4) unsuccessful validation data returned from any unsuccessful validation at the stage of accepting wine products. As wine products being processed and handled by different nodes along the supply chain with data updated by scanning NFC tags of the wine products using tag-reading *ScanWINE* with the state of wine record to be updated accordingly to the database. These categories of data are updated along the supply chain until the point of purchase at which wine consumers use tag-reading *ScanWINE* to scan NFC tags and retrieve data such as the wine pedigree data and transaction data for real-time validation to determine if a wine product is counterfeit or not.

A unique identifier is assigned to each wine product and is written into the NFC tag. Such tag-attached wine products are then shipped from winemakers to different nodes along the supply chain. During the transportation process of these wine products along the supply chain, each involved node could scan the NFC tags and adds the aforementioned four categories of data into these NFC tags, respectively. In this way, the next node can check whether or not the wine products have already passed through the legitimate supply chain. If any inconsistency is found at any node, such wine products may be considered as counterfeits and should be returned to winemakers. However, once the wine product reached post-purchase stage and circulated in any customer-to-customer markets, its authenticity is no longer guaranteed, as anyone who has an NFC reader can interfere and clone tags' data. Therefore, it is important to develop anti-counterfeiting systems that could work even when the data stored in tags is cloned in post-purchase supply chain with attacks detected and prevented on any potential adversary changes.

5. Security Analysis on NAS

In this section, a series of security analyses are performed on NAS, with findings of these security analyses also elaborated and discussed so as to identify which areas of NAS could be improved and strengthened, in terms of security, with the conception of decentralized product anti-counterfeiting and traceability solution enabled by blockchain technology.

5.1. Asset Analysis

The asset analysis of NAS, as described in Appendix B.1, lists ten constituent assets of NAS which are categorized into components of hardware, software, and data. Hardware assets are NFC-enabled mobile devices and NFC tags, while software assets of NAS are the two NFC-enabled mobile applications and the database operation web application. As described in the system overview of NAS, the data model of NAS is indeed based on four types of data assets: data of, wine pedigree, transaction history, supply chain, and unsuccessful records.

The (1) *NFC tags (A04)*, (2) *transaction history data (A06)* of a wine record, and the (3) *backend database (A10)*, which is solely managed by winemakers, storing all sorts of data components listed in the asset analysis, are among the three most probable system components susceptible to security risks according to ratings assigned to each component of confidentiality, integrity and availability (the CIA methodology).

5.2. Threat Analysis

Based on the result of asset analysis and system components identified, which are susceptible to security risks, a threat analysis is therefore performed against NAS, with every threat as described in Appendix B.2. Every possible threat to NAS could be categorized into either (1) physical NFC tag threats or (2) system threats.

Regarding physical NFC tag threats, the most probable and risky threats identified are as follows:

1. *Tag cloning (T01)* for which each NFC tag used for product tagging and anti-counterfeiting purpose has a unique identifier. If the identifier information is exposed to attackers, the data stored in a tag could easily be cloned into another tag.
2. *Tag disabling (T03)* for which adversaries could take advantage of wireless nature of NFC systems in order to disable tags, from any further interaction with NFC tags, temporarily or permanently, by changing the state of NFC tags.
3. *Tag's data modification (T04)* for which NFC tags use writable memory and so an adversary can take advantage of such a feature to modify or delete valuable data from memory of any involved NFC tag, during any tag reading and tag writing process. Unsecured configuration or even misconfiguration on NFC tags could also allow attackers compromising NAS as a whole.

These physical NFC tag threats are primarily attributed by a weak and fully centralized authentication and authorization mechanism adopted to change data states stored in NFC tags and any unsecured configuration on NFC tags during every tag writing and tag reading process.

While regarding system threats, the most probable and risky threats identified are as follows:

1. *Man-in-the-middle relay attack (T08)* for which in a relay attack, an adversary acts as a man-in-the-middle. An adversarial device is placed surreptitiously between a legitimate NFC tag and mobile applications or mobile applications with dedicated backend database which is on logged-in state. Adversaries could obtain unauthorized access or unintentional information disclosure on the confidential data related to transactions or the supply chain as a whole.
2. *Tracking and tracing (T10)*, for which, via sending queries and obtaining same responses from an NFC tag at various locations, it can then determine where an NFC tag of a specific product is located physically with location data supplied. A malicious hacker may also obtain login data, via brute-forcing or dictionary attacks, to spoof and log in to mobile applications or web applications registered with the same login data as legitimate users.
3. *Denial-of-service (T11)* for which DoS attacks are usually physical attacks, such as jamming the system with noise interference, blocking radio signals, removing, or even disabling NFC tags, causing different system components or the entire system to work improperly. Without sufficient auditing logs and monitoring functionalities of different system components, it would be unable to investigate any system misuse and compromise, security breach over leakage of confidential data.
4. *Spoofing attack on data of product records (T14)* when attackers get some information about the identity of NFC tags either by detecting communication between mobile applications and legitimate NFC tags or by physical exploration on these NFC tags, attackers could then clone the NFC tags. Poor code quality on microservices enabling interaction between mobile applications and the backend database of product records, such as weak authorization and authentication with dummy passwords or without database backup path, could lead to data theft on the confidential data of processed product records maintained in NAS.

These system threats are majorly attributed by the single-point processing, storage, and failure due to the fact that operations and data of NAS are managed and controlled solely by winemakers as the anti-counterfeiting and traceability features of NAS are built on specific winemakers and around industrial operations enabled by NFC technology or other tag-communication technologies. Furthermore, vulnerabilities such as weak authentication and authorization as well as in lack of sufficient auditing logs and effective API monitoring tools could also give rise to threats, such as man-in-the-middle relay attack, tracking-and-tracing, and spoofing attack. Adversaries could manipulate vulnerabilities to obtain unauthorized access or unintentional information disclosure over the confidential data such as the transaction data, wine pedigree data, or supply chain data. The disclosure of confidential data would then lead to adverse manipulation on product records and so disability of anti-counterfeiting functionalities of NAS could be expected. Adversaries could also make use of vulnerabilities, such as unsecured configuration on servers, poor security implementation over the code base on possible attacks, as well as lacking audit logs and API performance monitoring, to perform denial-of-service (DoS) attacks on different system components of NAS affecting its service availability, stability, and performance.

6. Discussion of Research Results

With blockchain fundamentals explained, how blockchain technology could impose security upgrades to existing product anti-counterfeiting and traceability systems, such as NAS, of supply chain industry, and the results gathered from security analyses performed on NAS, this section will cover the summary of vulnerabilities identified in existing product anti-counterfeiting and traceability systems. The opportunities of decentralizing anti-counterfeiting and traceability in supply chain industry and potential concerns on developing decentralized solutions for supply chain industry, are also identified.

6.1. Summary of Vulnerabilities on Centralized System Architecture

Among NAS and other existing anti-counterfeiting alternatives with centralized architecture, utilizing wireless tag communication technologies, there could be at least three common probable counterfeit attacks applied to these anti-counterfeiting solutions. These attacks manipulating threats listed under the *physical NFC tag threats* and *system threats* according to the threat analyses performed are (1) modification of product records stored in tags, such as fabricating product identifiers or vintages of any product; (2) cloning of metadata stored in tags such as those genuine product records to any counterfeit product tag; and (3) removal of a legitimate tag from a genuine product and its reapplication to any other counterfeit products.

It has come to a point that even though the implementation of NAS itself is already more effective and secured than most of its typical supply chain anti-counterfeiting and traceability counterparts, with original product records being validated at any node along the supply chain, the centralized architecture of NAS could still pose risks in data integrity and product authenticity as any node, not only winemakers, along the supply chain have full control of product records stored in their own database architectures. In case different nodes along the supply chain are untrusting to each other, there could still be possibilities that a product record being duplicated adversely leading to a situation that product consumers could still purchase a product counterfeit at retail points, with fabricated wine records retrieved from NAS or its counterparts implemented in specific supply chain industries.

The typical architecture of centralized supply chains creates several concerns. First, there is a tremendous processing burden on servers, as significant numbers of products processed by multiple supply chain nodes. Second, substantial storage is required to store authentication records for every single processed product. Third, with centralized systems, traditional supply chains inherently have the problem of single-point failure and so potential service downtime and data loss could be expected. All in all, centralized product anti-counterfeiting and traceability systems, such as NAS, rely on a centralized authority to combat counterfeit products which would result in *single-point processing, storage, and failure* and those potential attacks via manipulating the security threats identified in threat analysis performed against NAS.

6.2. Opportunities of Decentralizing Supply Chain Anti-Counterfeiting and Traceability

To better prevent risks and overcome threats with vulnerabilities initiated by centralized architecture, Blockchain Technology (or other Distributed Ledger Technologies built with decentralized networks) stands out as a potential framework to establish a modernized, decentralized, trustworthy, accountable, transparent, and secured supply chain innovation against counterfeiting attacks, compared with those developed on centralized architecture, with comparison between two as detailed in Figure 8.

Decentralized Architecture with Blockchain	Centralized Architecture
New data is only added when consensus is reached	New data is added via administrator without any consensus reached
Can only insert new data, old data is immutable	No restriction on any data modifications
Distributed in nature	Single point of failure
Decentralized in nature	Single point of control
Peer-to-peer structure with cloud instances	Client-server architecture
Cryptographic verification	Server performs actions on users' behalf
Resiliency and availability increased with number of peers	Backups and contingency plans are manually implemented
Cryptographic authentication and authorization	Cryptography implemented separately as add-ons

Figure 8. Comparison of decentralized and centralized architecture.

Given a variety of advantages, such as prevention of single-point failure, better resilience, and availability of being applied among supply chain participants, introduced with the blockchain technology and concept of decentralized application, to have a more secured and sophisticated supply chain system against counterfeiting attacks, it has well proven that decentralized supply chain anti-counterfeiting and traceability are in demand. The decentralized solutions are worth developing and implementing in supply chain industry, starting with a new solution or developing a novel prototype with a decentralized architecture, based on legacy solutions, such as NAS, to reinforce the innovative idea of product anti-counterfeiting. There are also a variety of opportunities that autonomous and decentralized supply chain anti-counterfeiting and traceability solutions could bring to supply chain industry as explained in the following.

6.2.1. Improved Data Integrity of Supply Chain

With the advent of blockchain technology and other technologies such as distributed file storage, a multi-layered data storage and validation mechanism, involved with on-chain and off-chain data operations, on product records could be implemented in decentralized solutions of supply chain anti-counterfeiting and traceability.

The on-chain and off-chain storage and validation could then be in place to ensure data integrity and prevent the decentralized solutions from any attempted attacks. The attacks could be cloning attacks on NFC tags, modification attacks in case product identifiers and signatures stored in NFC tags are inconsistent with its counterparts stored in the backend databases or on-chain storage with deployed smart contracts, or reapplication attacks in case both read count and write count are inconsistent to its counterparts stored off-chain and on-chain respectively.

With the decentralization enabled by the blockchain network, data integrity of processed product records is further improved which could further coupled with a concept of digital-asset tokenization representing every single product processed on the decentralized solutions. Any state change on specific product record could only be initiated with invoking methods on an open smart contract deployed to the network, with its transaction now needing to be validated with consensus reached, with other nodes held and run by different participating entities of supply chain industry, on the network. The immutability of transaction states related to any state transition of specific product record operations would mean that any state change processed on the network could be referred and queried based on individual transaction hashes and block numbers, which could further be confirmed on blockchain explorers connected to blockchain nodes running on the specific blockchain network.

6.2.2. Strengthened Security Considerations

Given the security considerations deployed to different possible validation mechanisms of any product record validation operation to improve data integrity, and according to the findings from the threat analysis performed against NAS, a variety of security attacks existed in NAS are no longer valid. Those attacks could well be prevented with errors thrown if they are detected before a state transition could be completed on a specific product record.

Any state transition on product records is required on-chain and off-chain validations performed against the data of product records, with respective transactions also validated by other blockchain nodes running on the same blockchain network. These validation steps of product record validation operation are now required to include signature generation and signing procedures, with key management modules offered to users, on any attempted state transition on product records. Regarding security considerations applied to the deployed smart contracts, which will be required if any blockchain 2.0 implementation is adopted to decentralize the supply chain anti-counterfeiting and traceability, multiple validation syntax on specific conditions are developed and included in different methods of smart contracts. This would prevent the system from being manipulated by potential attacks, such as reapplication attacks in which the on-chain write count and its counterpart stored off-chain do not actually match.

Design patterns with the role restriction concept of the deployed smart contract could also be introduced, so as to enable access authorization for authenticated node accounts held by specific entities of supply chain industry, to different methods defined in the deployed smart contracts. The security model on data integrity of NAS is currently based only on operations before the point of purchase. It could further be extended when the supply chain anti-counterfeiting and traceability functions are properly decentralized, as long as the post-purchase wine consumers of consumer-to-consumer market are also registered entities or even running nodes in the decentralized solutions.

6.2.3. High Availability of System Functionalities

Ensuring high-level operational performance of different system components to maintain system functionalities for different product record operations is key to the system implementation of NAS or any other supply chain anti-counterfeiting and traceability systems. With opportunities of system security and data integrity on product record now highly dependent on the system decentralization, the availability of data and states stored, as well as those decentralized system components, becomes more significant to the overall availability of system functionalities.

Regarding the decentralized solutions, the availability and resilience on data and states stored on blockchain network or any other decentralized system components are assured and can even be enhanced with increasing number of blockchain nodes running on the blockchain network owing to the fact that each node of these networks keeps the copy of the states stored in persistent volume dedicated to these distributed nodes. Availability of the blockchain network would also be enhanced with increased amount of blockchain nodes running on the blockchain network in which availability could be preserved as long as there is at least a blockchain node running on the network.

Dedicated persistent volume storage could also be assigned to each node instance running on the blockchain network to store blockchain states and individual chain data, so as to benefit from faster synchronization and data recovery on states to any new blockchain nodes connected to the network. This will then assure the availability of blockchain nodes and the blockchain network as a whole, as failed blockchain nodes could be reconnected to synchronize and process transactions sent to the network immediately. Nonetheless, availability of the data stored on-chain will be preserved with the smart contracts deployed to the blockchain network as long as it is actively running and mining new blocks constantly. The off-chain database and the app-backend service could also be made distributed with individual instances with which every participant under the decentralized solutions could now host their own instance of the supply chain anti-counterfeiting and traceability ecosystems.

6.3. Potential Concerns on Development of Decentralized Solutions

According to the summary of vulnerabilities on centralized system architecture, with opportunities of developing the decentralized solutions also identified, a set of fundamental system requirements of a decentralized version of NAS, namely, the Decentralized NFC-Enabled Anti-Counterfeiting System (dNAS), is also proposed with potential concerns on developing such decentralized solutions elaborated in the following.

6.3.1. Manageable System Integration Model

It is understood that not every user of the decentralized solutions would possess with in-house technical capacity to maintain its own instance of system components constituting the decentralized solutions. A manageable system integration model will need to be in place to help promote adoptions of the decentralized solutions from its centralized counterparts, which have been implementing and adopted in the wider supply chain industry, among different potential users in supply chain industry and for different stakeholders of industries to collaborate for good to help improve the worsening situation of product counterfeits, with the process integrity also conserved.

Such a manageable system integration model could provide another layer of indirection, allowing potential users to safely manage their own keys and the backup of product records in case anything unexpected goes wrong with system components of the decentralized solutions. An organization or alliance of major industry participants could act or be voted as leaders and host system components such as microservices, mobile applications, and even the decentralized blockchain nodes forming a blockchain network, as a fail-over and manage requests from these potential users of the decentralized solutions. A common data model of product records, applied to the decentralized solutions, should also be defined with additional metadata added after completing different types of industrial operations or steps declared in product data operation such as data validation and data creation steps both off-chain and on-chain, in order to facilitate a seamless process of data migration and integration.

6.3.2. Degree of Decentralization

The degree of decentralization is dependent and based on the chosen mechanism of manageable system integration. This will also require promoting adoption of such decentralized solutions, implying that some software components of the decentralized solutions are still expected to be hosted by intermediaries. The intermediaries could be backend databases, mobile applications, and backend application, with users only keeping hold of secrets or instances of decentralized system components such as blockchain nodes assigned to registered entities to the decentralized solutions.

The decentralized model could be substantiated with the blockchain network and its chosen consensus algorithm. Implementations of a blockchain network with its dedicated blockchain interface provide a certain degree of decentralization when it comes to transactions being validated and packed in a block on-chain with the respective methods of deployed smart contract invoked as well as proposing state changes on-chain registry for peer authentication and the blockchain network. Decentralization around smart contract management and deployment, management of software component instances of the decentralized solutions, and management of the distributed network protocols might be limited depending on the choice of blockchain implementations adopted but would definitely not be solely managed by product producers in the supply chain industry.

6.3.3. Limited Scalability

Scalability can somehow contradict with the degree of decentralization defined in the current setting of blockchain implementations. The decentralized solutions generally take more computation time for the same set of operations performed with centralized solutions, such as NAS, and so it is expected that decentralized systems are generally less scalable than its centralized counterparts owing to the fact that consensus is needed for every state changed on the data of processed product records.

The multi-layered validation and creation mechanism of product records processed in the decentralized solutions could in all likelihood imply decreased scalability as they could now involve more steps to store or even update representations of any product record, given these state transitions on product records involved are required to be processed in the decentralized system components as well. In case the number and size of product records grow with more products circulating in the supply chain, longer computation time is definitely needed for these product record operations with more computation resources committed on data processing of product records stored in the off-chain backend database structure. Data stored and processed on decentralized system components, such as the blockchain network, should be kept minimum, given the computation resources needed for these decentralized processes could be exponential compared with its centralized counterparts. The requests sent to the existing microservices and blockchain interface could be handled sequentially for which a new request could be processed only if the previous one is completed. The single-threaded handling of these microservices, involved in any end-to-end product record operations, could possibly hinder the scalability of the decentralized solutions as a whole.

Given the extra decentralized processes required in the decentralized solutions from the point of invoking endpoints made available in blockchain interface all the way to the blockchain network. The size of the blockchain will grow over time with more transactions validated and packed in blocks mined, not to mention a new block could be created in an interval of block time defined when initiating the blockchain network. It could take fairly long period of time for any new blockchain node to synchronize with other blockchain nodes to get to the latest global states of the network. The long synchronization time would hinder the user experience for participants using the decentralized solutions when the size of blockchain is too bulk. Though there are potential scalability concerns on any proposed decentralized solutions, it is still worth decentralizing the supply chain anti-counterfeiting and traceability, if comparing with the benefits brought by decentralization, such as the strengthened data integrity and improved system security with distributed instances enabling individual nodes along the supply chain collaborating to combat product counterfeits.

6.3.4. Potential Security Vulnerabilities

Every registered instance of the decentralized solutions is normally assigned with a blockchain node and an account of which a key pair could be generated and assigned to registered instances for their storage and management. Key pairs are required to validate and send transactions with its local blockchain node via the chosen blockchain client protocol. The same key pair could be retrieved and utilized over and over again without a concept of proper key rotation and management, and it is possible that the key pair could be compromised and thus the aforementioned attacks could still be made possible to create vulnerabilities and threats to the decentralized solutions.

Following the compromised key secrets, *distributed denial of service* (DDoS) could also be made possible as long as the blockchain node is hosted with enough computation resource, and it is possible to spam the blockchain network with a huge amount of transactions to be processed. The denial-of-service attack could also be performed by any malicious registered node, though they are part of the supply chain industry. A distributed key management module with key rotation functionalities to store and manage the key secrets should be implemented in the decentralized solutions, and a security authentication layer should also be added on top of the key management module whenever the blockchain interface, owned by the same registered instance, retrieves the key pair.

While for the data of product records stored and processed on decentralized components of decentralized solutions, such as the smart contracts deployed to the blockchain network and the service instance of any distributed file storage implemented, any blockchain node assigned to the registered entities could interpret and retrieve product record subsets if requested. The retrieval of product records is possible only if the unique identifier of processed products is supplied, as these product record subsets stored with the deployed smart contracts might not be encrypted and obfuscated with any hash algorithms. Though application of data security is really dependent on which types of blockchain network and consensus algorithm are adopted, which could generally categorized either into *"permissioned"* or *"permissionless"*, the decentralized solutions could adopt. Any data stored and processed on any decentralized system component should be kept minimum, obfuscated, and even encrypted to prevent from any potential malicious manipulation.

The publicity of the smart contract source code could also cause security vulnerabilities. Unlike the source code of different system components in NAS, which has the option to have its code base open-sourced or completely privatized, the smart contract code of decentralized solutions is always open and easily accessible by blockchain nodes running on the same network. Malicious users of the decentralized solutions running blockchain nodes could therefore look for human-induced vulnerabilities if any method of the deployed smart contracts is not implemented correctly.

6.3.5. Privacy Concern

As discussed in the potential security vulnerabilities, lacking a key rotation mechanism would imply the same public address is possible to be mapped to an actual registered node. The system role identifier of each registered node could further be mapped to a true identity of the representative organization or entity of the supply chain industry, by other registered nodes running on the same blockchain network if the decentralized solutions are developed in a setting with *permissioned blockchain implementations*.

Although public addresses stored on-chain could be obfuscated with hash functions applied, events could be emitted when methods of the deployed smart contracts are invoked, whenever there is a new transaction initiated on product record operations related to the same public addresses. The related events are later received by the event listener of every blockchain service instance. With more events emitted involving the same set of public addresses, it is more likely a specific public address could be mapped to an actual registered instance, and so its transaction volume could still be derived by other users of the decentralized solutions, which could potentially be its competitors. In addition to public addresses, these data fields could directly relate to physical entities and cause privacy concern if there is no privacy-preserving technology in place for these sensitive data fields. If the NFC tags are not deactivated properly when the respective products are consumed or transferred, it could possibly lead to a privacy threat based on any unencrypted or unobfuscated data field of specific product records stored in the NFC tags. Privacy-preserving technologies are required with use cases defined, based on chosen mechanisms on data processing and validation procedures to be included in the decentralized solutions.

7. Conclusions

Based on the growing concern in product counterfeiting of supply chain industry, the contributions of this research are to determine if anti-counterfeiting and traceability solutions currently adopted in supply chain industry could benefit from a certain degree of decentralization, via a series of security analyses: functional analyses performed against NAS. The possible opportunities and a list of fundamental system requirements of a decentralized version on NAS are also identified. Further statistical analyses on the research references of this research were also performed and demonstrated in Appendix A, where 86% of the research references of this research were published between 2012 and 2021, in which 44% were actually published within the past 5 years, and 20% of research references were actually published by IEEE.

A thorough explanation of blockchain technology is covered in this exploratory research so as to rationalize why the decentralized solutions should be developed with existing Blockchain 2.0 implementations, and why blockchain technology should be utilized to decentralize current solutions. There are mainly two categories of threats identified in these centralized solutions: *physical NFC tag threats* and *system threats*. These indicate the fact that tag-cloning, tag-disabling, tag's data modification, man-in-the-middle relay attack, spoofing attack on product data, and denial-of-service are among the most probable and risky threats to NAS as well as other existing anti-counterfeiting and traceability systems built with any software system component and wireless tag component.

The discussion of research results indicated some potential opportunities, which are exactly the potential advantages of decentralizing the existing supply chain anti-counterfeiting solutions where it has been maintained merely by product producers or in the case of NAS—the winemakers. The opportunities, such as (1) the improved data integrity with a decentralized product record management with state transitions on product records only accepted if its respective transaction is validated and the block is mined on the blockchain network; (2) the strengthen security considerations on data of product records processed with multilayered validations in place; and (3) the high availability on system components and data of product records enabled by the decentralized system components, including the blockchain network to prevent from possible system downtime and unavailability.

There are also potential concerns on actual development of the decentralized solutions: (1) the manageable system integration model on whether it could promote adoptions among the industry participants; (2) degree of decentralization which could be contradicting the performance and throughput of processing product records; (3) the limited scalability attributed by the extra decentralized processes and multilayered validations on processing product data; (4) the security vulnerabilities attributed by lack of proper key management and rotation mechanism; and (5) the privacy concern on true identities of targeted entities being exposed, which could be mapped basing on a public address of repetitive uses, by other malicious entities running instances of the decentralized solutions, and even nodes on the same blockchain network. A fundamental set of system requirements for the *Decentralized NFC-Enabled Anti-Counterfeiting System (dNAS)*, which is proposed to decentralize NAS of centralized architecture, is therefore defined, according to the potential concerns discussed. The system requirements defined will clarify the actual development and implementation of dNAS to take place in a separate research where the proposed system use cases, architecture and implementation will be elaborated with explanation on why the proposed decentralized solution could combat product counterfeiting in supply chain industry.

Funding: This research received no external funding.

Data Availability Statement: Not Applicable, the study does not report any data.

Conflicts of Interest: The author declares no conflict of interest.

Abbreviations

The following abbreviations are used in this manuscript:

NAS	The NFC-Enabled Anti-Counterfeiting System
NFC	Near-Field Communication
OECD	Organization for Economic Cooperation and Development
EUIPO	European Union Intellectual Property Office
POS	Point-of-Sale
RFID	Radio-frequency Identification
IoT	Internet of Things
PID	Product Identifier
dNAS	The Decentralized NFC-Enabled Anti-Counterfeiting System
PoW	Proof-of-Work Consensus Algorithm
UTXO	Unspent Transaction Outputs
EOA	Externally Owned Accounts
EVM	Ethereum Virtual Machine
Geth	Go-Ethereum Client
DoS	Denial-of-Service
GHOST	Greedy Heaviest Observed Subtree
ASIC	Application-Specific Integrated Circuit
GPU	Graphics Processing Unit
ÐApps	Decentralized Applications
CIA	Confidentiality, Integrity, and Availability

Appendix A. Statistical Analyses on Research References

The research references listed in this research are categorized by year of publication, as depicted in Figure A1, where 86% of the research references were published between 2012 and 2021.

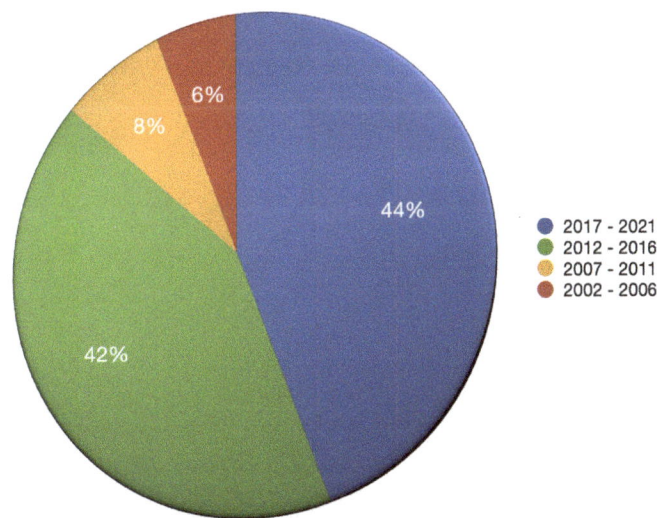

Figure A1. Research References By Publication Year.

The research references listed in this research are categorized by publisher, as depicted in Figure A2, where 20%, 13%, and 9% of the research references were published by IEEE, Springer, and Elsevier, respectively.

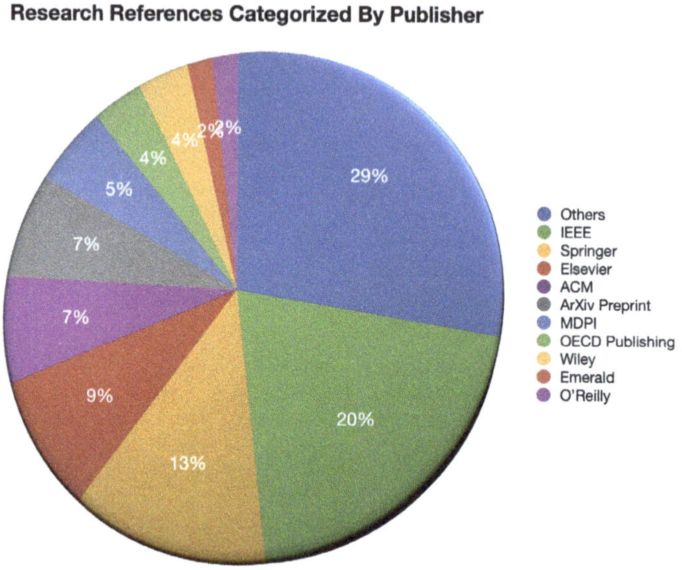

Research References Categorized By Publisher

- Others: 29%
- IEEE: 20%
- Springer: 13%
- Elsevier: 9%
- ACM: 7%
- ArXiv Preprint: 7%
- MDPI: 5%
- OECD Publishing: 4%
- Wiley: 4%
- Emerald: 2%
- O'Reilly: 2%

Figure A2. Research references by publisher.

Appendix B. Security Analysis on NAS

Appendix B.1. Asset Analysis of NAS

The asset analysis of NFC-Enabled Anti-Counterfeiting System (NAS) is demonstrated in Figure A3.

ID	Asset Type	Description	Confidentiality	Integrity	Availability
A01	Mobile Devices	The mobile device the users need to use and interact with the tags and the wine products	MEDIUM	MEDIUM	MEDIUM
A02	Mobile Application – TagWINE	The mobile application to write data to the tags of the wine products	MEDIUM	HIGH	HIGH
A03	Mobile Application – ScanWINE	The mobile application to read data from the tags of the wine products	MEDIUM	HIGH	HIGH
A04	NFC Tags	The NFC tags attached on the wine products storing the data	HIGH	HIGH	HIGH
A05	Wine Pedigree Data	The data of the wine products	MEDIUM	HIGH	HIGH
A06	Transaction History Data	The transaction data of a specific wine product based on its unique identifier	HIGH	HIGH	HIGH
A07	Supply Chain Data	The supply chain data of a specific wine product based on its unique identifier	MEDIUM	HIGH	HIGH
A08	Unsuccessful Record Data	The data of the unsuccessful validation of a specific wine product	MEDIUM	HIGH	MEDIUM
A09	Web-based Application	The user interface for users to manage the inventory of the wine products they hold	MEDIUM	HIGH	HIGH
A10	Backend Database	The database of all the data stored, based on the schema of different sorts of data as mentioned	HIGH	HIGH	HIGH

Figure A3. Asset analysis of NAS.

Appendix B.2. Threat Analysis of NAS

The threat analysis of NFC-Enabled Anti-Counterfeiting System (NAS) is demonstrated in Figure A4.

ID	Type	Threat	Vulnerability	Assets Affected	Impact	Likelihood	Risk
T01	Physical NFC Tag Threats	Tags Cloning	Weak authentication on reading or writing the data in NFC tags	A04, A05, A06, A07	Each NFC tag used for product tagging and anti-counterfeiting purpose has a unique identifier. If the identifier information is exposed by attackers, the data stored in tags can easily be copied.	HIGH	HIGH
T02		Reversed Engineering	Weak authentication on reading or writing the data in NFC tags; Lack of tamper-resistant mechanism applied to NFC tags	A04, A05, A06, A07, A10	An attacker takes apart the chip to find out how it works in order to receive data from the Integrated Circuit (IC) because most NFC tags are not equipped with a tamper resistant mechanism for an estimated long period of time.	MEDIUM	HIGH
T03		Tags Disabling	Weak authentication; Weak authorization	A04, A05, A06, A07	Adversaries could take advantage of the wireless nature of NFC systems in order to disable tags temporarily or permanently by changing the state of these tags.	HIGH	HIGH
T04		Tags' Data Modification	Unsecured configuration on NFC tags; Weak authentication	A04, A05, A06, A07, A10	NFC tags use writable memory and so an adversary can take advantage of this feature to modify or delete valuable data from memory of any tag. Unsecured configuration or misconfiguration on NFC tags could also allow attackers to compromise the service system as a whole.	HIGH	HIGH
T05		Power Analysis	Weak authorization; Unsecured configuration on NFC tags; Lack of sufficient auditing log	A04	Power analysis attacks can be mounted on NFC systems by monitoring the power consumption levels of NFC tags.	MEDIUM	LOW
T06	System Threats	Phishing	Weak authentication; Weak authorization; Lack of sufficient auditing log	A01-A04, A09	Phishing attacks could easily be performed when the NFC tags are modified or replaced with other tags. Phishing in NFC tag could be done by social engineering. Attackers could try to mislead NFC users by social engineering. If the tags are altered, it is easy to deceive the users of both NFC-enabled applications to reveal their personal information by misleading them into malicious applications or websites.	MEDIUM	MEDIUM
T07		Replay Attack	Accessible communication; Weak authentication; Single-point processing	A04-A08, A10	The replay attack is when a malicious node or device replays that key information, which is eavesdropped through the communication between different system components of NAS. Although passive interference is usually unintentional, an adversary can take advantage of the data gathered from compromised components and replay any tag-writing or tag-reading process which could change the state of wine records.	MEDIUM	HIGH
T08		Man-in-the-middle Relay Attack	Weak authentication; Weak authorization; Single-point processing	A02-A09	In a relay attack, an adversary acts as a man-in-the-middle. An adversarial device is placed surreptitiously between a legitimate NFC tag and mobile applications or mobile applications and backend database with the logged-in state. Adversaries could obtain unauthorized access or manipulate unintentional information disclosure over the confidential data.	HIGH	HIGH
T09		Eavesdropping	Accessible communication; Weak authentication; Weak authorization; Poor security implementation over the codebase on attack; Lack of API performance monitoring	A05-A08	An unauthorized use of mobile applications could eavesdrop conversations between or interact with, without owner's knowledge, a tag and mobile device with those mobile application to obtain account data, location data, wine product data or transaction data leading to unintentional information disclosure.	HIGH	MEDIUM
T10		Tracking and Tracing	Weak authentication; Weak authorization	A05-A10	By sending queries and obtaining the same response from a tag at various locations, it can be determined where the specific tag of the wine product physically is and which locations it has visited. A malicious hacker may also gain login details, via brute-forcing or dictionary attacks, to spoof and login to mobile application and web application registered with the same email address as legitimate users.	HIGH	HIGH
T11	System Threats	Denial of Service	Unsecured configuration on servers; Poor security implementation over the codebase on attack; Lack of sufficient auditing log; Lack of API performance monitoring	A02-A03, A09-A10	DoS attacks are usually physical attacks like jamming the system with noise interference, blocking radio signals, or even removing or disabling NFC tags, causing the system to work improperly. Without sufficient auditing logs and monitoring functionalities of different system components or the system as a whole, would not enable investigation on any system misuse and compromise, security breach over leakage of confidential data.	HIGH	HIGH
T12		Viruses	Lack of antivirus design pattern to the codebase or antivirus software applied; Lack of sufficient auditing log	A01-A04, A09-A10	Virus could be a serious threat to an NFC system. A virus programmed on an NFC tag by an unknown source could cripple an NFC system whenever the tagged item is read at a facility. Service system and device could then be prone to virus/worm infection leading to service unavailability or corruption on confidential data collected, processed and shared.	MEDIUM	HIGH
T13		Crypto Attacks	Lack of proper source control and configuration management; Poor security implementation over the codebase on attack; Weak authentication; Weak authorization	A02-A08, A10	Since most NFC systems use encryption technology to ensure the confidentiality and integrity of the data delivery, attacking against the encryption algorithm is a common form of attack and also lack of code review and access control of the database and codebase may lead to a situation that hackers could steal the collected data, analytics or codebase not yet published or intended to share.	MEDIUM	HIGH
T14		Spoofing Attack on Data of Wine Records	Poor security implementation over the database on caches on the server; Lack of sufficient auditing log	A02-A10	When attackers get some information about the NFC tags either by detecting communication between mobile applications and legitimate NFC tags or by physical exploration of the tags, attackers can clone the tags. Poor code quality on microservices enabled mobile applications interacting with the database, such as dummy password or without database backup path, could lead to data theft on confidential data.	HIGH	HIGH

Figure A4. Threat analysis of NAS.

Appendix C. Proposed System Requirement of dNAS

Fundamental System Requirements of dNAS

The system requirements of the proposed Decentralized NFC-Enabled Anti-Counterfeiting System (dNAS) are demonstrated in Figure A5.

ID	Title	Rationale	Description
R01	Use of Blockchain Technology	Advantages of Decentralized Architecture	dNAS shall use distributed ledger technology to cryptographically validate data of wine products at different nodes along the supply chain.
R02	Autonomous Data Validation	Availability of Data Validation	dNAS shall use product unique identifiers to validate data of specific wine products via the NFC tags and the state of product records at any point along the supply chain.
R03	Decentralized Consensus	Avoidance of Single-point Processing	Each node along the supply chain should provide updated supply chain data and transaction data of specific wine products in order for that product to be traced and validated, with consensus made amongst the registered nodes of supply chain, which are also running nodes on the blockchain network.
R04	Confidentiality of Trading Volumes of Registered Nodes	Data Privacy on State of Nodes	dNAS should maintain confidentiality of trading volumes between the nodes along the supply chain and also nodes of a potential consortium with a running blockchain network.
R05	Data Integrity Verification	Data Integrity	dNAS must provide a mechanism to verify integrity of individual data points listed in the predefined record schema of wine products.
R06	Authentication on Registered Nodes	Improved Authentication	dNAS shall authenticate registered nodes, including its respective authorised device and application account, both on-chain and off-chain.
R07	Authorization on Any State Change	Authorization on Functionalities	Only registered nodes which are also authorized by winemakers of wine products or a consortium administrator shall have the ability to append or update data fields of transferred wine products.
R08	Data Invisibility	Data Integrity	dNAS must prevent any unauthorized node from directly viewing the actual data stored on-chain or off-chain.
R09	Invisibility of Transaction Senders	Privacy of Registered Nodes	dNAS shall prevent anyone from linking transactions with any registered nodes running on the blockchain network.
R10	Limited Node Registry	Blockchain Integrity	dNAS must prevent unapproved nodes from running on the blockchain network.
R11	Expected Scalability – Data Record Volume	Scalable dNAS	dNAS should be able to at least take in _1-million records_ of wine products without significantly affecting other operational parameters.
R12	Expected Scalability – Transaction Per Second	Scalable dNAS	The blockchain network of dNAS should be able to process at least _500 transactions per second_.
R13	New Registration On-Boarding	Ease of Node Registration	The blockchain architecture of dNAS should provide a mechanism for new instances to join the ecosystem without negatively affecting other nodes or the integrity of the architecture.
R14	Notification of State Changes	State Transparency	dNAS should have a mechanism to inform the registered nodes of state changes of a specific wine product.
R15	Choice of Consensus	Consensus Performance	dNAS should utilize a blockchain network with the choice of consensus algorithm and its consensus reached in a decentralized fashion, least round of communication before consensus is reached, most performing, and easiest to be run by registered nodes.
R16	Registry Voting	Ease of Node Registration	dNAS should provide a mechanism to register nodes in a decentralized way and once the registration is reached, the newly registered nodes will be allowed to become operational on dNAS.
R17	Smart Contract Upgradeability	Updated Decentralized Code	dNAS should provide a mechanism for deployed smart contracts to be upgraded.
R18	Blockchain Interface	Ease of Integration	Blockchain smart contracts must be accessible from an instance of any registered node via defined RESTful APIs in the blockchain interface.
R19	Permissions of State-Transitioned Functions	Improved Authentication and Authorization	Only permissioned accounts shall be allowed to invoke to specific methods of any deployed smart contract. Everyone else shall not be allowed to do this.

Figure A5. System requirements of proposed dNAS.

References

1. OECD; EUIPO. *Trade in Counterfeit and Pirated Goods–Mapping The Economic Impact*; OECD Publishing: Paris, France, 2019.
2. Dégardin, K.; Roggo, Y.; Margot, P. Understanding and fighting the medicine counterfeit market. *J. Pharm. Biomed. Anal.* **2014**, *87*, 167–175. [CrossRef] [PubMed]
3. OECD; EUIPO. *Trade in Counterfeit Goods and Free Trade Zones–Evidence From Recent Trends*; OECD Publishing: Paris, France, 2018.
4. Holz, H.J.; Applin, A.; Haberman, B.; Joyce, D.; Purchase, H.; Reed, C. Research Methods in Computing: What are they, and how should we teach them? In Proceedings of the Working Group Reports on ITiCSE on Innovation and Technology in Computer Science Education, Bologna, Italy, 26–28 June 2006; pp. 96–114.
5. Chowdhury, B.; Chowdhury, M.; Abawajy, J. Securing a smart anti-counterfeit web application. *J. Netw.* **2014**, *9*, 2925–2933. [CrossRef]
6. Ilic, A.; Lehtonen, M.; Michahelles, F.; Fleisch, E. Synchronized secrets approach for RFID-enabled anti-counterfeiting. In Proceedings of the Demo at Internet of Things Conference, Zurich, Switzerland, 26–28 March 2008.
7. Cheung, H.; Choi, S. Implementation issues in RFID-based anti-counterfeiting systems. *Comput. Ind.* **2011**, *62*, 708–718. [CrossRef]
8. Lee, W.H.; Chou, C.M.; Wang, S.W. An NFC Anti-Counterfeiting framework for ID verification and image protection. *Mob. Netw. Appl.* **2016**, *21*, 646–655. [CrossRef]
9. Wazid, M.; Das, A.K.; Khan, M.K.; Al-Ghaiheb, A.A.D.; Kumar, N.; Vasilakos, A.V. Secure authentication scheme for medicine anti-counterfeiting system in IoT environment. *IEEE Internet Things J.* **2017**, *4*, 1634–1646. [CrossRef]
10. Han, S.; Bae, H.J.; Kim, J.; Shin, S.; Choi, S.E.; Lee, S.H.; Kwon, S.; Park, W. Lithographically encoded polymer microtaggant using high-capacity and error-correctable QR code for anti-counterfeiting of drugs. *Adv. Mater.* **2012**, *24*, 5924–5929. [CrossRef]
11. You, M.; Lin, M.; Wang, S.; Wang, X.; Zhang, G.; Hong, Y.; Dong, Y.; Jin, G.; Xu, F. Three-dimensional quick response code based on inkjet printing of upconversion fluorescent nanoparticles for drug anti-counterfeiting. *Nanoscale* **2016**, *8*, 10096–10104. [CrossRef] [PubMed]
12. Koh, R.; Schuster, E.W.; Chackrabarti, I.; Bellman, A. *Securing the Pharmaceutical Supply Chain*; White Paper; Auto-ID Labs, Massachusetts Institute of Technology: Cambridge, MA, USA, 2003; Volume 1, p. 19.
13. Shi, J.; Li, Y.; He, W.; Sim, D. SecTTS: A secure track & trace system for RFID-enabled supply chains. *Comput. Ind.* **2012**, *63*, 574–585.
14. Staake, T.; Michahelles, F.; Fleisch, E.; Williams, J.R.; Min, H.; Cole, P.H.; Lee, S.G.; McFarlane, D.; Murai, J. Anti-counterfeiting and supply chain security. In *Networked RFID Systems and Lightweight Cryptography*; Springer: Berlin/Heidelberg, Germany, 2008; pp. 33–43.
15. Staake, T.; Thiesse, F.; Fleisch, E. Extending the EPC network: The potential of RFID in anti-counterfeiting. In Proceedings of the 2005 ACM Symposium on Applied Computing, Santa Fe, NM, USA, 13–17 March 2005; pp. 1607–1612.
16. Kwok, S.; Ting, S.; Tsang, A.H.; Cheung, C. A counterfeit network analyzer based on RFID and EPC. *Ind. Manag. Data Syst.* **2010**, *110*, 1018–1037. [CrossRef]
17. Aung, M.M.; Chang, Y.S. Traceability in a food supply chain: Safety and quality perspectives. *Food Control* **2014**, *39*, 172–184. [CrossRef]
18. Nakamoto, S. Bitcoin: A Peer-to-Peer Electronic Cash System. Available online: https://bitcoin.org/bitcoin.pdf (accessed on 26 March 2021).
19. Gupta, S.; Sadoghi, M. Blockchain Transaction Processing. Available online: https://expolab.org/papers/BC-txn-Encyclopedia.pdf (accessed on 26 March 2021).
20. Buterin, V. Ethereum white paper. *GitHub Repos.* **2013**, *1*, 22–23.
21. Androulaki, E.; Barger, A.; Bortnikov, V.; Cachin, C.; Christidis, K.; De Caro, A.; Enyeart, D.; Ferris, C.; Laventman, G.; Manevich, Y.; et al. Hyperledger fabric: A distributed operating system for permissioned blockchains. In Proceedings of the Thirteenth EuroSys Conference, Porto, Portugal, 23–26 April 2018; pp. 1–15.
22. Zheng, Z.; Xie, S.; Dai, H.; Chen, X.; Wang, H. An overview of blockchain technology: Architecture, consensus, and future trends. In Proceedings of the 2017 IEEE International Congress on Big Data (BigData Congress), Honolulu, HI, USA, 25–30 June 2017.
23. Peters, G.W.; Panayi, E. Understanding modern banking ledgers through blockchain technologies: Future of transaction processing and smart contracts on the internet of money. In *Banking Beyond Banks and Money*; Springer: Berlin/Heidelberg, Germany, 2016; pp. 239–278.
24. Gervais, A.; Karame, G.O.; Wüst, K.; Glykantzis, V.; Ritzdorf, H.; Capkun, S. On the security and performance of proof of work blockchains. In Proceedings of the 2016 ACM SIGSAC Conference on Computer and Communications Security, Vienna, Austria, 24–26 October 2016; pp. 3–16.
25. Meynkhard, A. Fair market value of bitcoin: Halving effect. *Investig. Manag. Financ. Innov.* **2019**, *16*, 72–85. [CrossRef]
26. Rosenfeld, M. Analysis of hashrate-based double spending. *arXiv* **2014**, arXiv:1402.2009.
27. Haferkorn, M.; Diaz, J.M.Q. Seasonality and interconnectivity within cryptocurrencies-an analysis on the basis of bitcoin, litecoin and namecoin. In *International Workshop on Enterprise Applications and Services in the Finance Industry*; Springer: Berlin/Heidelberg, Germany, 2014; pp. 106–120.
28. Wood, G. Ethereum: A secure decentralised generalised transaction ledger. *Ethereum Proj. Yellow Pap.* **2014**, *151*, 1–32.

29. Frantz, C.K.; Nowostawski, M. From institutions to code: Towards automated generation of smart contracts. In Proceedings of the 2016 IEEE 1st International Workshops on Foundations and Applications of Self* Systems (FAS* W), Augsburg, Germany, 12–16 September 2016; IEEE: Piscataway, NJ, USA, 2016; pp. 210–215.
30. Rouhani, S.; Deters, R. Security, performance, and applications of smart contracts: A systematic survey. *IEEE Access* **2019**, *7*, 50759–50779. [CrossRef]
31. Bartoletti, M.; Pompianu, L. An empirical analysis of smart contracts: Platforms, applications, and design patterns. In *International Conference on Financial Cryptography and Data Security*; Springer: Berlin/Heidelberg, Germany, 2017; pp. 494–509.
32. Dannen, C. *Introducing Ethereum and Solidity*; Springer: Berlin/Heidelberg, Germany, 2017; Volume 318.
33. Benet, J. Ipfs-content addressed, versioned, p2p file system. *arXiv* **2014**, arXiv:1407.3561.
34. Swan, M. *Blockchain: Blueprint for a New Economy*; O'Reilly Media, Inc.: Newton, MA, USA, 2015.
35. Crosby, M.; Pattanayak, P.; Verma, S.; Kalyanaraman, V. Blockchain technology: Beyond bitcoin. *Appl. Innov.* **2016**, *2*, 71.
36. Bahack, L. Theoretical bitcoin attacks with less than half of the computational power (draft). *arXiv* **2013**, arXiv:1312.7013.
37. Kwon, J. Tendermint: Consensus without Mining. Available online: https://cdn.relayto.com/media/files/LPgoWO18TCeMIggJVakt_tendermint.pdf (accessed on 26 March 2021).
38. Buchman, E. Tendermint: Byzantine Fault Tolerance in the Age of Blockchains. Ph.D. Thesis, University of Guelph, Guelph, ON, Canada, 2016.
39. Cheng, J.C.; Lee, N.Y.; Chi, C.; Chen, Y.H. Blockchain and smart contract for digital certificate. In Proceedings of the 2018 IEEE International Conference on Applied System Invention (ICASI), Tokyo, Japan, 13–17 April 2018.
40. Zyskind, G.; Nathan, O. Decentralizing privacy: Using blockchain to protect personal data. In Proceedings of the 2015 IEEE Security and Privacy Workshops, San Jose, CA, USA, 21–22 May 2015.
41. Casino, F.; Dasaklis, T.K.; Patsakis, C. A systematic literature review of blockchain-based applications: Current status, classification and open issues. *Telemat. Inform.* **2019**, *36*, 55–81. [CrossRef]
42. Al-Jaroodi, J.; Mohamed, N. Blockchain in industries: A survey. *IEEE Access* **2019**, *7*, 36500–36515. [CrossRef]
43. Khezr, S.; Moniruzzaman, M.; Yassine, A.; Benlamri, R. Blockchain technology in healthcare: A comprehensive review and directions for future research. *Appl. Sci.* **2019**, *9*, 1736. [CrossRef]
44. Drosatos, G.; Kaldoudi, E. Blockchain applications in the biomedical domain: A scoping review. *Comput. Struct. Biotechnol. J.* **2019**, *17*, 229–240. [CrossRef] [PubMed]
45. Abeyratne, S.A.; Monfared, R.P. Blockchain ready manufacturing supply chain using distributed ledger. *Int. J. Res. Eng. Technol.* **2016**, *5*, 1–10.
46. Caro, M.P.; Ali, M.S.; Vecchio, M.; Giaffreda, R. Blockchain-based traceability in Agri-Food supply chain management: A practical implementation. In Proceedings of the 2018 IoT Vertical and Topical Summit on Agriculture-Tuscany (IOT Tuscany), Tuscany, Italy, 8–9 May 2018.
47. Toyoda, K.; Mathiopoulos, P.T.; Sasase, I.; Ohtsuki, T. A novel blockchain-based product ownership management system (POMS) for anti-counterfeits in the post supply chain. *IEEE Access* **2017**, *5*, 17465–17477. [CrossRef]
48. Kennedy, Z.C.; Stephenson, D.E.; Christ, J.F.; Pope, T.R.; Arey, B.W.; Barrett, C.A.; Warner, M.G. Enhanced anti-counterfeiting measures for additive manufacturing: Coupling lanthanide nanomaterial chemical signatures with blockchain technology. *J. Mater. Chem. C* **2017**, *5*, 9570–9578. [CrossRef]
49. Kim, H.M.; Laskowski, M. Toward an ontology-driven blockchain design for supply-chain provenance. *Intell. Syst. Account. Financ. Manag.* **2018**, *25*, 18–27. [CrossRef]
50. Singh, R.; Dwivedi, A.D.; Srivastava, G. Internet of things based blockchain for temperature monitoring and counterfeit pharmaceutical prevention. *Sensors* **2020**, *20*, 3951. [CrossRef]
51. Yiu, N.C.K. An NFC-Enabled Anti-Counterfeiting System for Wine Industry. *arXiv* **2014**, arXiv:1601.06372.

Short Biography of Authors

Neo C. K. Yiu is a computer scientist and software architect specialized in developing decentralized and distributed software solutions for industries. Neo is currently the Lead Software Architect of Blockchain and Cryptography Development at De Beers Group on their end-to-end traceability projects across different value chains with the Tracr™ initiative. Formerly acting as the Director of Technology Development at Oxford Blockchain Society, Neo is currently a board member of the global blockchain advisory board at EC-Council. Neo received his MSc in Computer Science from University of Oxford and BEng in Logistics Engineering and Global Supply Chain Management from The University of Hong Kong.

MDPI
St. Alban-Anlage 66
4052 Basel
Switzerland
Tel. +41 61 683 77 34
Fax +41 61 302 89 18
www.mdpi.com

Future Internet Editorial Office
E-mail: futureinternet@mdpi.com
www.mdpi.com/journal/futureinternet

www.ingramcontent.com/pod-product-compliance
Lightning Source LLC
LaVergne TN
LVHW070654100526
838202LV00013B/959